A MECHANISTIC APPROACH TO **PLANKTON ECOLOGY**

A MECHANISTIC APPROACH
TO **PLANKTON ECOLOGY**

Thomas Kiørboe

PRINCETON UNIVERSITY PRESS • PRINCETON AND OXFORD

Copyright © 2008 by Princeton University Press

Published by Princeton University Press, 41 William Street, Princeton, New Jersey 08540

In the United Kingdom: Princeton University Press, 6 Oxford Street, Woodstock, Oxfordshire OX20 1TW

Library of Congress Cataloging-in-Publication Data

Kiørboe, Thomas.
A mechanistic approach to plankton ecology / Thomas Kiørboe.
    p. cm.
Includes bibliographical references and index.
ISBN 978-0-691-13422-2 (alk. paper)
   1. Plankton—Ecology.   I. Title.
   QH90.8.P5K56 2008
   578.77'6—dc22

                               2007048545

British Library Cataloging-in-Publication Data is available

This book has been composed in Utopia Typeface

Printed on acid-free paper. ∞

press.princeton.edu

Printed in the United States of America

10 9 8 7 6 5 4 3 2 1

CONTENTS

ILLUSTRATIONS

TABLES

PREFACE

A N EARLIER VERSION of the first five chapters of this book was put together during the winter of 2003/2004 for a graduate course in Biological Oceanography that I was teaching at the University of Southern Denmark. Chapter 7 was added in January 2006 for a post-graduate course in Barcelona, Spain, and the remaining chapters were written in the spring of 2007, at which time the earlier material was also updated and streamlined.

Much of the material in this book has been developed in collaboration with students, postdocs, and colleagues through discussions and joint research projects, as will be evident from the quoted references. I thank them all. Special thanks go to George Jackson, Andy Visser, and Uffe Thygesen for many fruitful discussions, for sharing their ideas, and in particular for offering their expertise and help in mathematics and physics where mine was lacking.

Charlottenlund, Denmark
March 2007

A MECHANISTIC APPROACH TO **PLANKTON ECOLOGY**

Chapter One

INTRODUCTION

1.1 BIOLOGICAL OCEANOGRAPHY—MARINE BIOLOGY—OCEAN ECOLOGY

I LIKE SPORTS FISHING. I used to have a small boat from which I fished for herring, cod, and trout in the Øresund. Angling there can be quite productive, in particular if you learn where to go. At certain spots there are more cod than at others, for whatever reason. Once such spots were found, we used to find them again from land or sea marks. For example, 200 m north of a particular green buoy very often there would be cod. Eventually I knew so many "hot spots" that, whenever I went fishing, I would catch some cod. And eventually this lost its excitement. I sold the boat and started fishing for trout from the coast. Here the approach is not blind. You can see both the fish and its environment. You learn to know where to find the fish using relevant cues—that is, not a green buoy but a certain type of vegetation or structure of the seafloor. You learn to think as a trout; you develop intuition. When you visit new beaches, and if you are good enough, you can "read" the coast, and you can find fish. You can extrapolate the insights gained at one site to new, unknown locations.

Much of our knowledge of the biology of the oceans is derived from "blind" sampling. We use instruments to measure bulk properties of the environment, such as salinity and temperature, and we use bottle or net samples to extract knowledge about the organisms living in the ocean. This kind of approach has contributed important knowledge but has also influenced the way we view marine life. It leads us to focus on abundances, production rates, and distribution patterns. Such a perspective is very relevant in the context of the ocean as a resource for fisheries. It is also helpful in developing an understanding of biogeochemical issues such as ocean carbon fluxes. But on its own, this approach is insufficient, even for those purposes. The kind of intuition that we develop about marine life is, of course, influenced by the way we observe it, and because the ocean is inaccessible to us, and most planktonic organisms are microscopic, our intuition is rudimentary compared, for example, to the intuitive understanding we have about (macroscopic) terrestrial life. Our understanding of the biology of planktonic organisms is still based mainly on examinations of (dead) individuals, field samples, and incubation experiments, and even our sampling may be severely biased toward those organisms that are not destroyed by our harsh sampling methods. Similarly, experimental observations are limited to those organisms that we can collect live and keep and cultivate in the laboratory. One may argue that these limitations have biased our understanding of the ecology of the plankton and thus constrained our comprehension of the function of pelagic food webs (Smetacek and Pollehne 1986).

The ocean is structured on all spatial scales, but sampling averages over volumes, and this has led us to focus our attention on the potential importance of certain scales over others. In particular, sampling averages over volumes that exceed the ambit of individual plankters by orders of magnitude. Thus, we may relate organism distributions to distributions of salinity and temperature, for example, without knowing whether these are the cues to which the organisms respond. From sampling we can enumerate phytoplankton, zooplankters, and other particles, but we do not know how they are distributed relative to one another at a scale that is relevant to the organisms. We do not know whether the particles in a sample were originally aggregated as marine snow because such aggregates disintegrate to component particles when sampled by traditional means. In a water sample, we can also measure the concentrations of various solutes (dissolved organics, oxygen, nutrient salts), which may lead us to think that the solutes were homogeneously distributed in the sample before it was collected; they rarely are. And we may incubate water samples with radiotracers to get estimates of production rates of bacteria and phytoplankton. The implicit assumption is that such rates are representative of the corresponding rates in situ. But are they? These numbers, concentrations, and rates that are measured at scales that exceed the daily ambit of the individual organisms by orders of magnitude are insufficient to provide an understanding of how the organisms function in their environment. Visualization and observations as well as considerations at the level of the individual plankter are keys to establishing a mechanistic understanding of how the organisms function and interact and, hence, how the system of which they are part works (e.g., Azam and Long 2001).

Ecosystems consist of populations, which in turn consist of individuals that interact with one another and with the environment. Biological interactions in the ocean are not between populations or between trophic levels, as many box-model representations of pelagic food webs might lead us to think. Trophic levels and populations are abstractions, and interactions occur at the level of the individual. "Blind" sampling of bulk properties may result in observed distributional patterns, for example, that cannot be

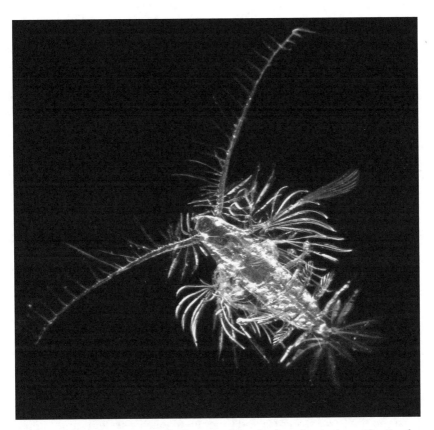

Fig. 1.1. *Haloptilus plumosa.* The beauty of the plankton is a main motivation for studying it. The copepod in the picture has extensive setation on all appendages that together forms a large basket. While feeding, a feeding current keeps the animal stationary in the water, ventral side up. Sinking particles and particles drawn in by the feeding current are collected by the large basket. Courtesy of Russ Hopcroft.

understood and explained from such an approach on its own. The picture must be complemented by approaches that consider the individual in its immediate environment and that provide a mechanistic understanding of the functioning of individuals and of components of the larger systems. This allows us to build models and to extrapolate observations beyond the system in which the observations were made.

Traditionally, scientists who go on cruises and examine distribution patterns of both biota and environmental properties using sampling are considered biological oceanographers, and those who explore the functioning of individuals, for example by conducting laboratory experiments with organisms, are considered marine biologists. We need to combine the two approaches to understand the ecology of the oceans. This book considers the functional ecology of planktonic organisms but with a view on how organism biology shapes the ecology of the oceans. In fact, as we proceed, and particularly by the end, we shall try to integrate our understanding of individual-level processes to predict properties of planktonic populations and of pelagic food webs.

The motivation to try to understand the ecology of planktonic organisms is twofold. The first driving force has to do with a simple interest in natural history. It is fascinating to watch the behavior of live plankters under the microscope or—better—free-swimming plankters by video; they have different but often beautiful forms and colors, and even closely related species may behave very differently, which makes identifying live plankton much easier than identifying dead ones. My own love for the plankton was elicited by one photograph in particular, of a very beautiful echinoderm larva, published in a popular book by Gunnar Thorson, and by the beauty of plankton in general (fig. 1.1). This fascination led me to a desire to understand why and how a plankter does what it does and how it solves various problems, among them, finding food in a dilute environment. The second reason for examining the adaptations and behavior of plankters is our interest in understanding overall properties of pelagic systems and how the pelagic system relates to the larger-scale issues of fisheries' yield, CO_2 balance, global climate, and others. Understanding the mechanics of individual behaviors and interactions may allow us to predict rates and to scale rates to sizes, which, in turn, may help us understand the (size) structure and function of pelagic systems and to predict effects of environmental changes and human impacts.

1.2 THE ENCOUNTER PROBLEM

Life is all about encounters. In the ocean, for example, phytoplankton cells need to encounter molecules of nutrient salts and inorganic carbon; bacteria need to encounter organic molecules; viruses need to encounter their hosts; predators need to encounter their prey; and males need to encounter females (or vice versa). There is no life without encounters, and the pace of life is intimately related to the rate at which encounters happen. Other important processes in the ocean, such as the formation of marine snow aggregates, likewise depend on encounters, here encounters between the component particles. We are therefore interested in encounter rates. And we are interested in understanding the mechanisms that govern encounters. What are the constraints, and what are the implications of these constraints?

All organisms, including plankters, have three main tasks in life, namely to eat, to reproduce, and to avoid being eaten, all related to encounters or avoiding encounters. The behavior, morphology, and ecology of planktonic organisms must to a large extent represent adaptations to undertake these missions, and the diversity of form, function, and behavior that we can observe among plankters must be the result of different ways of solving the problems in the environment in which they live. The pelagic environment seen from the point of view of a small plankter is very different from the environment experienced by humans, and our intuition is often insufficient to allow us to understand the behavioral adaptations of planktonic organisms. Thus, although ornithologists to a large extent may be able to understand the behavior of their study organisms by using common sense, copepodologists rarely can, to rephrase the title of a classical ecology paper (Hutchinson 1951). For example, at the scale of planktonic organisms, the medium is viscous, and inertial forces therefore are insignificant, which makes moving an entirely different undertaking than what we as humans are used to or have seen other terrestrial animals do; the density of water is orders of magnitude higher than the density of air, which makes floatation easier and currents more important; for the smallest pelagic organisms (bacteria), thermally driven Brownian motion makes steering impossible; and most plankton use senses different from, and less far-reaching than, vision to perceive the environment. In addition, the pelagic environment is three-dimensional, whereas humans mainly move in only two dimensions. This implies, among other things, that average distances between a planktonic organism and its target may be very large, maybe thousands of body lengths. Because of the often nonintuitive nature of the immediate environment of small pelagic organisms, we need to appeal to fluid dynamic considerations in order to achieve a mechanistic understanding of the small-scale interactions between plankters and their environment.

In pursuing the encounter problem we can write a very general equation that describes encounter rates

$$E = \beta C_1 C_2 \tag{1.1a}$$

where E is the number of encounters happening per unit time and volume between particle types 1 and 2, C_1 and C_2 are the concentrations of these particles, and β is the *encounter rate kernel* ($L^3 T^{-1}$) (see table 1.1 for definition of symbols used). Often we are interested in looking at the per capita rate, that is, the rate at which one particle of type 1 encounters a particle of type 2:

$$e = E/C_1 = \beta C_2 \tag{1.1b}$$

TABLE 1.1
Definitions of Symbols Used

Symbol	Definition	Dimensions
a	Radius	L
$\lvert\mathbf{a}\rvert$	Acceleration	LT^{-2}
C	Concentration	ML^{-3} or L^{-3}
d	Separation distance	L
D	Diffusion coefficient	L^2T^{-1}
E	Encounter rate	$L^{-3}T^{-1}$
e	Specific encounter rate	T^{-1}
f	Production rate of fertile eggs	T^{-1}
I	Light intensity	$L^{-2}T^{-1}$
J	Flux	$MT^{-1}L^{-2}$
k	Specific light attenuation coefficient	L^2
K	Boltzmann's constant	$ML^{-2}T^{-1}\,{}^{\circ}K^{-1}$
l_x	Survivorship at age x	—
L	Plume length	L
m_x	Fecundity at age x	T^{-1}
M	Carrying capacity	L^{-3}
Pe	Péclet number	—
Q	Flow (of molecules or particles)	MT^{-1}
r	Radial distance	L
R	Reaction distance	L
R_0	Net reproductive rate	—
Re	Reynolds number	—
S	Signal strength	LT^{-1}
Sh	Sherwood number	—
t	Time	T
T	Generation time	T
U, u	Speed	LT^{-1}
x,y,z	Distance along $x-$, y, or $z-$axis	L
α	Stickiness	—
β	Encounter rate kernel	L^3T^{-1}
γ	Shear rate	T^{-1}
δ	Step length	L
Δ	Deformation rate	T^{-1}
ε	Turbulent dissipation rate	L^2T^{-3}
ζ	Egg–hatching time	T
η	Kolmogorov length scale	L
ι	Handling time	T
κ	Maturation age	T
λ	Detachment rate	T^{-1}
ς	Dynamic viscosity	$ML^{-1}T^{-1}$
μ	Specific growth rate	T^{-1}
ν	Kinematic viscosity	L^2T^{-1}

TABLE 1.1 (*continued*)

Symbol	Definition	Dimensions
ρ	Density	ML^{-3}
σ	Mortality	T^{-1}
τ	Run duration	T
Φ	Volume fraction	—
ω	Vorticity	T^{-1}

The dimensions are L for length, T for time, M for mass, and °K for degree Kelvin.

For example, if particle 1 is a suspension-feeding ciliate and C_2 the concentration of its phytoplankton prey, then β is the ciliate's clearance rate, and e its ingestion rate (assuming that all encountered particles are ingested). The clearance rate is the equivalent volume of water from which the ciliate removes all prey particles per unit time. In many suspension-feeding ciliates, the clearance rate can be interpreted directly as a filtration rate; that is, the rate at which water is passed through a filtering structure that retains suspended particles. As a different but similar example: if particle 1 is a fish larva looking for food, and particle 2 its microzooplankton prey, then β is the volume of water that the larvae can search for prey items per unit time; if all encountered prey are consumed, then e is the ingestion rate of the fish larva. We may also see the process from the point of view of the prey, in which case βC_1 is the mortality rate of the phytoplankton or microzooplankton prey population through ciliate grazing or fish larval feeding. As a final example: if C_1 is the concentration of bacteria, and C_2 the concentration of organic molecules on which the bacteria feed, then e is the assimilation rate; it is more difficult to give a physical interpretation of β in this case. However, it is, like a clearance rate, the imaginary volume of water from which the bacterium removes all molecules per unit time. In fact, any encounter problem that I can think of can be cast in terms of the general equation (eq. 1), but obviously the interpretation or meaning of the terms may be very different.

The processes or mechanisms responsible for encounters are contained in the encounter-rate kernel. Obviously, from the examples above, these mechanisms are diverse. Intuitively, encounter rates depend on two factors: the motility of the encountering "particles" and the ambient fluid motion that may enhance encounter rates. Motility encompasses here the diffusivity of molecules, the swimming of organisms, and the sinking of particles. In regard to planktonic organisms, ambient fluid motion essentially means turbulence because planktonic organisms (contrary to benthic ones) are embedded in the general flow. From this consideration, one can see that there may be different components entering into the encounter-rate kernel depending on the specific problem under consideration.

The encounter problem—and equation 1—will be the primary issue throughout most of this book and of chapters 2–5 in particular. We shall examine the encounter problem for a number of fundamental processes in the plankton (e.g., feeding, mating, coagulation). And for each of the processes considered, we shall try to write simple models for β, as far as possible based on first principles. My presentation is case driven, but wherever possible or necessary, I will make excursions to more general explanations.

1.3 This Book

The purpose of this book is twofold. First and foremost, I want to explore the ecology of plankton organisms at the level of the individual by examining how they are adapted to the viscous, three-dimensional, and dilute (in terms of food) environment in which they live. My goal is to provide a mechanistic understanding of the functioning of individual plankters, both in terms of the interactions between the organisms and their immediate environment and in terms of the interactions between individuals. These are the topics of chapters 2–6. Second, I want to use the insights into individual-level processes to examine population- and ecosystem-level processes and patterns. Biological processes in the plankton occur at the level of the individual, and biological interactions in the ocean are between individuals rather than between populations or trophic levels; the latter are abstractions. Although the structure and function of pelagic food webs cannot be derived solely from a mechanistic understanding of the functioning of the individuals, important properties of distribution, population, and community patterns, and of the turnover of matter and energy in the plankton can be predicted from individual-level processes. This extension is the topic of chapters 7 (population processes) and 8 (pelagic food webs). The intention of these two final chapters is to illustrate the usefulness of combining oceanographic and biologic approaches to examine ocean ecology rather than to provide a complete description of pelagic ecosystems. Overall, throughout this book, emphasis will be on telling a coherent rather than a comprehensive story, and there are therefore many aspects of plankton ecology that are not covered here.

The intended audience of this book is motivated graduate and postgraduate students as well as researchers interested in plankton ecology. Most of the text has been developed for and used as teaching material for such groups. Knowledge of basic concepts of biological oceanography and plankton ecology is assumed. Many biologists are afraid of hard sciences, such as physics and mathematics. However, to understand the adaptations and ecology of small pelagic organisms, we need to draw on simple fluid dynamics and diffusion theory. Use of mathematics is, if not unavoidable, then at least extremely useful in describing and communicating such issues. We shall keep it as simple as possible, and high school level performance in calculus is sufficient to be able to follow the arguments, if not always all details in the less important and trivial derivations. This is not only for pedagogic reasons—I am not a mathematician myself and, hence, am largely restricted to high school level math. Often I shall refrain from formal proofs but rather copy solutions from books and papers where such proofs have been derived. Emphasis is on understanding the causal relations and the idea of the argument rather than the formal proofs.

There are several books that treat the same general topic of small-scale biological/physical interactions as this one and provide both a much broader and deeper level of insight. I have myself been much inspired by Denny (1993), Vogel (1994), and Berg (1993), and the last two must be considered classics. All three books are written at a level that biology students will be able to read. Okubo (1980) is another classic but somewhat more difficult to read. None of these books refers particularly to plankton, and the present volume is an attempt to apply the principles described in the above books to plankton ecology. I will make particular reference to Denny (1993), and there will be some overlap between his book and this one. Finally, chapter 2 of Mann and Lazier (1991) provides an easy and helpful introduction to several of the topics covered here.

I will give many examples, and much of this account will in fact be driven by examples from which I shall try to generalize. I must confess that the choice of examples is very biased toward my own research and that of my students and close colleagues. This is not because these examples are the best but mainly because these are the examples through which I myself developed my understanding of planktonic organisms and pelagic food webs.

Chapter Two

RANDOM WALK AND DIFFUSION

2.1 RANDOM WALK AND DIFFUSION

W E SHALL START by considering classical diffusion theory and by deriving Fick's first law. Many important and exciting phenomena in plankton biology are governed by molecular diffusion (e.g., nutrient acquisition in unicellular organisms) or can be analyzed using a random-walk diffusion analogy (e.g., organism motilities).

Solute molecules travel at very high speeds, tens or hundreds of meters per second, but because they constantly bump into water molecules, they do not get very far. Every time they collide with another molecule, their course is changed to a new random direction. This random walk is (Brownian) diffusion. The motility of many organisms can, as mentioned, also be described as a random walk. The classical example is the motility of bacteria. Many bacteria swim along more or less straight lines, interrupted only by "tumbles," where they stop and then continue swimming in a new, random direction. The similarity between molecular diffusion and the run–tumble motility of bacteria is striking. As we shall see later, even motilities of organisms that do not swim in a run-tumble mode may be described using the diffusion analogy.

The following is largely copied from Denny (1993), who in turn copied his account from the 1983 edition of Berg (1993):

Consider N particles moving only along the x-axis and all starting at $x=0$ (we shall later expand the one-dimensional case to two and three dimensions). Assume that every τ seconds, each particle will move one step of length δ to either the right or to the left (equally likely). Thus, the average speed of the particles is δ/τ. After n time steps, the position the ith particle is

$$x_i(n) = x_i(n-1) \pm \delta \qquad (2.1)$$

Note that the + or − sign will apply to approximately half of the particles each. Each particle will follow a different, random path (fig. 2.1A, B). The average position of all N particles can be obtained by averaging either the left or the right side of equation 2.1 and equating the two, i.e.,

$$\langle x(n) \rangle = \frac{1}{N} \sum_{i=1}^{N} x_i(n) = \frac{1}{N} \sum_{i=1}^{N} x_i(n-1) - \frac{1}{N} \sum_{i=1}^{N} \pm \delta$$

$$= \frac{1}{N} \sum_{i=1}^{N} x_i(n-1) = \langle x(n-1) \rangle \qquad (2.2)$$

because the term containing $\pm\delta$ vanishes. The triangular brackets, $\langle \rangle$, mean "average of." That is, on average, the particles do not move anywhere (fig. 2.1B).

But the particles of course do not all remain at $x=0$; they spread because they move (fig. 2.1B). To see how much the particles move, we may consider the average of the absolute values of their step lengths, or, as is the tradition, we may average the squared distances (these are all positive, of course) and then take the square root of the mean of the squares yielding the root-mean-square net distance traveled. This is the same as the standard deviation of the position of the particles. First consider the squared distances; by squaring both sides of equation 2.1 we get

$$x_i^2(n) = x_i^2(n-1) \pm 2\delta x_i(n-1) + \delta^2 \qquad (2.3)$$

then taking the mean of both the left and right side of equation 2.3

$$\langle x^2(n) \rangle = \frac{1}{N} \sum_{i=1}^{N} x_i^2(n-1) + \frac{1}{N} \sum_{i=1}^{N} \pm 2\delta x_i(n-1) + \frac{1}{N} \sum_{i=1}^{N} \delta^2$$

$$= \langle x^2(n-1) \rangle + \delta^2 \qquad (2.4)$$

That is, for each time step the mean square distance increases by δ^2. Because the mean square position at $t=0$ is 0 (i.e., $\langle x_i^2(0) \rangle = 0$), it follows that

$$\langle x^2(n) \rangle = n\delta^2 \qquad (2.5)$$

and the root-mean-square (*RMS*) distance traveled is, therefore,

$$\langle x^2(n) \rangle^{0.5} = \sqrt{n\delta^2} \qquad (2.6)$$

Recall that a step was taken every τ seconds; hence, the number of steps taken during time t since start is $n=t/\tau$. Therefore,

$$\langle x^2(t) \rangle^{0.5} = \sqrt{\frac{\delta^2}{\tau} t} \qquad (2.7)$$

We now define the diffusion coefficient

$$D = \delta^2/2\tau. \qquad (2.8)$$

The factor of ½ is introduced for "mathematical convenience." (I have never understood why this is convenient.) We now have

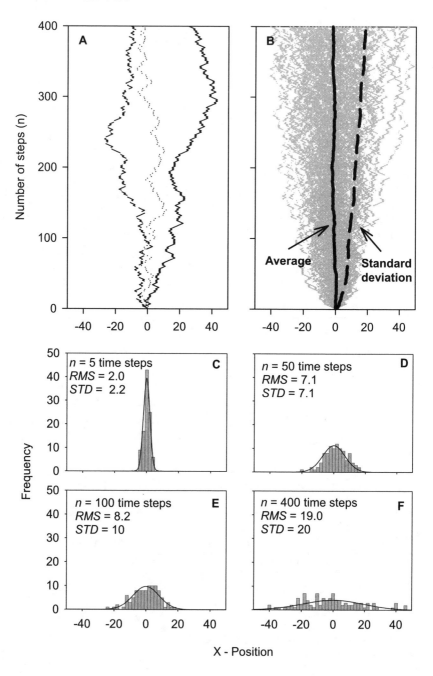

$$\langle x^2(t)\rangle^{0.5} = (2Dt)^{0.5} \tag{2.9}$$

What this means is that the *RMS* net distance traveled by the particles increases with the square root of time. In perhaps more familiar terms, the *RMS* net distance traveled at time t is the standard deviation of the distribution of the particles along the x-axis at time t (fig. 2.1B–F).

An important property of the above process is that despite the unpredictability of the direction of any single step, and of the path of any individual particle, the population of particles moves in a predictable way: the population will spread symmetrically from the starting position, and at any point in time it will be distributed approximately normally around the starting point with a standard deviation that increases with the square root of time (fig. 2.1C–F). In other words, out of random, individual movements arises diffusion that is a regular and predictable property of the population. This illustrates more generally how one is able to extrapolate individual processes to population processes over many orders of magnitude. We shall utilize this repeatedly in the rest of this book.

We can expand the one-dimensional considerations to two and three dimensions by noting that diffusion in the different dimensions is independent of one another. Thus, if the squared net distance traveled in the x-direction is $2Dt$, then it is the same in the y- and z-directions. Therefore, using Pythagoras, the *RMS* net distances traveled (r) in two and three dimensions are:

$$\langle r_{2d}^2 \rangle^{0.5} = (\langle x^2 \rangle + \langle y^2 \rangle)^{0.5} = (4Dt)^{0.5} \tag{2.10}$$

$$\langle r_{3d}^2 \rangle^{0.5} = (\langle x^2 \rangle + \langle y^2 \rangle + \langle z^2 \rangle)^{0.5} = (6Dt)^{0.5} \tag{2.11}$$

Thus, the factor of 2 in equations 2.8 and 2.9 is just replaced by 4 or 6 to described diffusion in two or three dimensions.

Fig. 2.1. Simulation of individual particle paths (A, B) and resulting distributions of 100 particles along the x-axis after $n=5$, 50, 100, and 400 time steps (C–F). Three particles follow unpredictable individual random paths (A), while the paths of 100 particles spread in a systematic way (B). The mean distance to the starting point of 100 particles remains constantly near 0, whereas the *RMS* net distance traveled (=standard deviation) increases with the number of time steps, approximately as $\sqrt{n}$. Distributions of the 100 particles at distinct times all approximate normal distributions (continuous curves in C–F) with mean 0 and standard deviation $STD = \sqrt{n}$. The standard deviations (*RMS*) of the simulated particle paths are similar, but not identical, to the expected standard deviations.

In order to develop our intuition, let us consider diffusion of biologically relevant molecules in water: oxygen, carbon dioxide, amino acids, or sugars. They all have diffusion coefficients of the order of 10^{-5} cm²s⁻¹. The time scale of diffusion in three dimensions is $t = RMS^2/6D$ (reorganize eq. 2.11). Thus, the time required for diffusion to transport molecules a distance (RMS) 1 μm is $(10^{-4})^2/(6 \times 10^{-5})$ s $\approx 2 \times 10^{-4}$ s, that is, very fast. Similarly, the time required to diffuse 1 mm is ~3 min, and 1 cm ~5 h. Thus, one can see that diffusion is a very rapid transport process at the small spatial scale but that it takes a very long time for substances to be transported by diffusion over longer distances—diffusion time increases with the square of the distance. The following page, http://www.bam.ie/bambrat/particle/particle.html, has a nice simulation of diffusing particles.

2.2 EXAMPLE: BACTERIAL MOTILITY

Before continuing with Fick's law, let us consider an example where we can apply our new insight and equations 2.8 and 2.9 (or their two-dimensional equivalents) to describe the motility of bacteria. As mentioned above, many bacteria swim in a run-tumble mode, and their motility may thus be described as diffusion and quantified by a diffusion coefficient. One can observe swimming bacteria under the microscope at low magnification using dark-field illumination, allowing one to videotape and subsequently digitize the two-dimensional projections of the swimming tracks (fig. 2.2). There is quite a diversity in swimming patterns among bacteria (many more examples can be found in Johansen et al. 2002). One of the species shown in figure 2.2 has relatively long runs between tumbles (strain HP11), another species tumbles much more frequently (HP46). Knowing the run length and swimming velocity, we can estimate diffusivities from equation 2.8, noting, however, that we are considering two dimensions rather than one (i.e., $D = \delta^2/4\tau$; the bacteria of course swim in all three dimensions, but we are looking only at the two-dimensional projection). Recall that the particle speed $u = \delta/\tau$; hence, $D = u^2\tau/4$. Here we interpret u as the average swimming velocity of the bacteria, and τ as the average run duration. This expression is correct only if all runs are of identical duration. In figure 2.2, this is obviously not the case. If tumbles, rather, occur at random intervals (as a Poisson process), and this is a sound and common assumption,[1] then run dura-

---

[1] There are other types of random walks where the run durations follow different probability distributions than the one assumed here. One such group of random walks, Lévy walks, assume a power law distribution of run lengths and lead to faster dispersion (superdiffusive; root-mean-square net displacement increases with time raised to a power

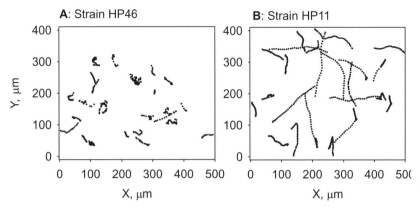

Fig. 2.2. Swimming tracks of two bacterial strains, both isolated from marine particles, HP46 (A) and HP11 (B) (from Kiørboe et al. 2002). The dots represent positions at 0.08- (A) or 0.16-s (B) intervals. Swimming patterns recorded in microscope preparation using dark-field illumination.

tions are exponentially distributed, and it can be shown that $D = u^2 \tau/2$ in two dimensions or, generally, $D = u^2 \tau/n$ in n dimensions (Visser and Thygesen 2003).

There is an additional factor that we must consider before we insert observed velocities and run lengths. It may be evident from figure 2.2 that the change in direction from run to run is biased; that is, all directions are not equally likely. It is very common among marine pelagic bacteria that directions are biased backward (e.g., Mitchell et al. 1996, Johansen et al. 2002), as is the case for HP11. Even though run directions are biased, they can be random in the sense that the direction of one run is independent of the direction of all preceding runs. It can be shown that in n dimensions (see Berg 1993 for a referenced derivation):

$$D = \frac{u^2 \tau}{n(1 - \alpha)} \tag{2.12}$$

larger than 0.5). Such random walks have been applied to describe the motility of microbes and other organisms (Viswanathan et al. 1996, Levandowsky et al. 1997, Bartumeus et al. 2003), but they are difficult to distinguish experimentally from diffusive random walks. Moreover, Lévy walk statistics require that the organism has "memory" in the form of knowledge of all former and future step lengths—an assumption that is difficult to accept. The applicability of Lévy walks to describe organism motility is therefore questionable (see also Edwards et al. 2007).

TABLE 2.1
Swimming Data for Two Strains of Bacteria

Strain	τ (s)	α	u (cm s^{-1})	D (cm^{2}s^{-1})
HP11	14.9	−0.67	42×10^{-4}	8×10^{-5}
HP46	0.4	0.48	46×10^{-4}	0.8×10^{-5}

Average run length, average dimensional swimming velocity, and directional bias (α) of tumbles as well as computed diffusivities for the two bacterial strains shown in Fig. 2.2. Data from Kiørboe et al. (2002) (corrected for errors).

where α is the mean value of the cosine of the angle between successive runs. Inserting observed run lengths, swimming speeds, and values of α in equation 2.12 yields estimates of D that are on the order of 10^{-5} cm^{2}s^{-1} (table 2.1).

Run-tumble motility, where the runs between tumbles are really linear, is strictly possible only for relatively large bacteria because small bacteria cannot steer. Thermally driven Brownian motion causes bacteria (and other particles) to rotate and, hence, to swim along nonstraight paths. The runs of the bacteria of strain HP11 appear a little wiggled, which may be a result of Brownian rotation (fig. 2.2B). The effect of rotation varies inversely with the particle radius cubed, and there is therefore a relatively sharp size limit of around 0.6 µm diameter below which swimming bacteria lose directional persistence during runs (Dusenbery 1996), and equation 2.12 becomes invalid. However, the combined effect of swimming, tumbling, and rotation still leads to a diffusive motility pattern. We may therefore alternatively estimate bacterial diffusivity using equation 2.10. The equation says that the root-mean-square distance increases with the square root of the time and that the proportionality constant is $2\sqrt{D}$. The expected square-root relationship comes out nicely for HP46 (fig. 2.3A), and a diffusivity can be computed from the estimated relationship ($D = 0.53 \times 10^{-5}$ cm^{2}s^{-1}), which is similar to that found above. However, for HP11, the observations do not fit the expected square-root relationship (fig. 2.3B). The *RMS* distance covered increases with time to the power of 0.8 rather than 0.5! The reason is that the diffusion analogy is valid only for $t \gg \tau$, and this assumption is violated here. The average run length of ~15 s in HP11 is longer than the time that a bacterium stays within the field of view in the microscope. We simply cannot follow the bacteria long enough, and at the smaller time and spatial scales at which we can observe the bacteria, the motility of HP11 cannot be described by the diffusion analogy. We shall later return to this problem of scale-dependent characterization and properties of motility patterns (chapter 4, section 4.10).

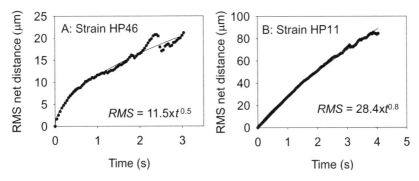

Fig. 2.3. Net displacement versus time in two bacteria. Root-mean-square net distance covered (*RMS*) as a function of time for the two bacterial isolates shown in Fig. 2.2. At the spatiotemporal scale considered, strain HP46 obeys the square-root dependence on time, but strain HP11 does not.

2.3 FICK'S FIRST LAW

We now return to the mathematical formalism based on the work of Denny (1993) and Berg (1993). We now know what diffusion and a random walk are. In many situations we need to apply Fick's laws, which describe the spatial and temporal distribution of particles that undergo random walks. Consider a situation in which we know the abundance of particles along the x-axis shown in figure 2.4. After one time step, τ, about half the particles at x, $N(x)$, will have crossed the dotted line and moved to $x+\delta$. Likewise, about half the particles at $x+\delta$, $N(x+\delta)$, will have crossed the dotted line in the opposite direction and have moved to position x. Thus, the net number of particles crossing the line in the x-direction is

$$-\frac{1}{2}\left[N(x+\delta)-N(x)\right] \tag{2.13}$$

The net flux of particles (J), that is, the net number passing per unit area and unit time, is obtained by dividing the net number crossing by the area (A) perpendicular to the x-axis and by the time interval, τ, hence:

$$J = -\frac{\left[N(x+\delta)-N(x)\right]}{2A\tau} = -\frac{\delta^2}{2\tau}\frac{1}{\delta}\left[\frac{N(x+\delta)}{A\delta}-\frac{N(x)}{A\delta}\right] \tag{2.14}$$

where the final expression was obtained by multiplying by δ^2/δ^2 and rearranging. Note that the first term in the last equation, $\delta^2/2\tau$, is the diffusion coefficient as defined above and that the terms in the square brackets are the particle concentrations, C, at positions $x+\delta$ and x because $A\delta$ is a volume. Hence:

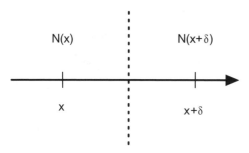

Fig. 2.4. Schematic for explaining Fick's first law. The number of particles at x is $N(x)$, and the number of particles at $x + \delta$ is $N(x + \delta)$. We are considering the number of particles passing the dotted line in either direction.

$$J = -D\frac{C(x + \delta) - C(x)}{\delta} \tag{2.15}$$

For $\delta \to 0$, we get

$$J = -D\frac{dC}{dx} \tag{2.16}$$

This is Fick's first law. It says that the net diffusive flux of particles is proportional to the concentration gradient. The minus sign indicates that the net transport is down the gradient, that is, from high to low concentration.

2.4 Diffusion to or from a Sphere

Concentration gradients arise at all spatial and temporal scales in the ocean and because of many different processes. Because diffusion is most efficient in transporting material over short distances, here we shall consider small-scale phenomena. A local sink of particles or molecules, such as an ambush predator eating all arriving prey particles, or an osmotroph absorbing all arriving nutrient molecules, will generate a local concentration gradient and, hence, cause a net flux of diffusing particles toward the collector. Envisage now a spherical collector of radius a surrounded by imaginary spherical shells of radius r (fig. 2.5).

The collector could, for example, be a phytoplankton cell taking up diffusing nutrient molecules. At steady state, the flow[2] of particles toward the collector through each of these shells must be the same; other-

---

[2]*Flow* and *flux*: the term flux is often used in a confusing sense. By *flux* we here always mean the net transport of particles/molecules *per unit area and unit time*, whereas the

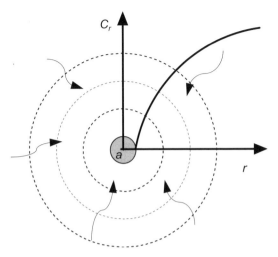

Fig. 2.5. Diffusion to sphere. A spherical collector of radius a surrounded by imaginary spherical shells of increasing radius. At steady state the number of molecules or particles passing each of these shells per unit time must be the same and equal to the rate at which molecules of particles are taken up by the collector; therefore, the flux and, hence, the concentration gradient varies inversely with the square of the distance to the center (eq. 2.18), and the concentration inversely with the distance (eq. 2.19). The term *collector* is generic for any organism or particle that collects molecules or particles.

wise, material would constantly accumulate or be depleted between shells, and we would not have steady state. This flow is identical to the collector's uptake rate, the rate at which a phytoplankton cell absorbs nutrient molecules. The flow (Q) can be estimated from Fick's first law by multiplying the flux by the surface area of the enveloping shells (recall that the surface area of a sphere is $4\pi r^2$):

$$Q = J 4\pi r^2 = -D\frac{dC}{dr}4\pi r^2 \qquad (2.17)$$

The concentration as a function of distance to the center of the sphere, C_r can now be found by rearranging equation 2.17

$$\frac{dC}{dr} = -\frac{Q}{4\pi D r^2} \qquad (2.18)$$

and integrating

---

flow is the flux integrated over a particular surface area. The dimensions of flux are $ML^{-2}T^{-1}$, and the dimensions of flow are MT^{-1}.

$$C_r = \int_{\infty}^{r} \frac{dC}{dr} dr + C_\infty = \int_{\infty}^{r} \frac{-Q}{4\pi Dr^2} dr + C_\infty = \frac{Q}{4\pi Dr} + C_\infty \tag{2.19}$$

where C_∞ is the concentration far away. If the concentration at the surface ($r=a$) is C_a, by replacing C_a with C_r and a with r in equation 2.19, we get

$$Q = 4\pi Da(C_a - C_\infty) \tag{2.20}$$

or, for a perfect absorber with $C_a = 0$:

$$Q = -4\pi DaC_\infty \tag{2.21}$$

The minus sign is annoying but correct and just says that the net flow is from high toward low concentration. If $C_a > C_\infty$ in equation 2.20, corresponding to a situation where the sphere is producing rather than absorbing material, the flux is away from the sphere. Note that equation 2.21 has the same form as our general encounter equation (eq. 1.1), with the specific encounter rate (e) given by $-Q$, and the encounter-rate kernel for diffusion given by

$$\beta_{\text{diffusion}} = 4\pi Da \tag{2.22}$$

2.5 FEEDING ON SOLUTES

Equations 2.20–2.22 are typically used to describe nutrient uptake in osmotrophs (e.g., phytoplankton cells taking up inorganic nutrients or bacteria taking up organic molecules), but they apply generally to encounters between any spherical collector and particles with a random-walk type of motility (see examples below). We first consider osmotrophs.

One immediate implication of equations 2.20–2.22 is that uptake rate scales with cell radius. For a spherical cell of volume $4/3\pi a^3$ this implies that the volume-specific uptake rate (Q') scales inversely with the second power of the radius:

$$Q' = \frac{Q}{\frac{4}{3}\pi a^3} = \frac{3DC_\infty}{a^2} \tag{2.23}$$

Thus, small cells are much more efficient than large cells in taking up nutrients. This is an old insight (Munk and Riley 1953), and it is consistent with most people's intuition, but for the wrong reason. Common wisdom argues that small cells have a larger surface area per volume than larger cells and, therefore, take up nutrients more efficiently. However, nutrient uptake in phytoplankton and pelagic bacteria is probably rarely limited by surface area; rather, it is limited by the rate at which diffusion can deliver molecules to the surface of the cells, which is the process considered here. Also, diffusion limitation is a much more severe constraint for large cells than surface area: the specific surface area of a cell (=surface area per volume) scales inversely with its radius, whereas the specific diffusive delivery of molecules scales inversely with its radius squared, as we saw above.

Large cells thus easily become diffusion limited and are, therefore, at a competitive disadvantage in oligotrophic oceans. This is probably the main reason that oligotrophic oceans are dominated by pico- and nano-phytoplankton, and large diatoms are restricted to more eutrophic conditions. We can illustrate this constraint by a simple calculation. Nitrogen is normally considered the limiting element for primary production, at least in coastal waters. During winter and early spring in coastal temperate waters, inorganic nitrogen occurs mainly in the form of nitrate, and in typical concentrations of 10^{-5} M. In the course of the spring, nitrate is depleted, and during summer, nitrogen is available mainly in the form of ammonium at a typical concentration of 10^{-7} M. If we assume that nitrogen uptake limits the growth rate of phytoplankton, then the specific growth rate, μ, must equal the specific uptake rate of nitrogen, $-Q'_N$ divided by the specific nitrogen content of the cells (C'_N=about 1.5 M):

$$\mu = -Q'_N / C'_N = \frac{3D_N C_{N,\infty}}{a^2 C'_N} \tag{2.24}$$

which we can solve for a, the radius of the cell:

$$a = \left(\frac{3D_N C_{N,\infty}}{\mu C'_N} \right)^{0.5} \tag{2.25}$$

If we assume as a first approximation that the phytoplankton growth rate is constantly 1 d^{-1} throughout the growth season (in fact it varies somewhat with the light intensity), then from the ambient nitrogen concentrations above, we can compute the maximum possible cell size from equation 2.25 (assuming $D=10^{-5}$ cm^{2}s^{-1}). During the spring period with high ambient nutrient conc entration, the maximum possible cell size (radius) is about 45 μm, whereas during summer it is 4.5 μm. (As an exercise, try to do the computation yourself—it is simple in principle, but be aware of the units!) The computed expected decrease in maximum cell size during the season largely fits with observations. The spring community in coastal temperate waters is dominated by large diatoms, whereas the summer community is characterized by small flagellates. Thus, the seasonal succession in phytoplankton composition is, at least in part, simply a result of constraints on nutrient acquisition. We shall return to this topic in chapter 8.

Equations 2.20–2.21 also explain why only very small cells such as bacteria can feed on dissolved organic material. To illustrate, we can use equation 2.21 to compute the maximum, diffusion-limited uptake rates of solutes. Bacteria take up amino acids, for example, and because nitrogen is often considered the limiting element for productivity in the ocean, we can to a first approximation consider the amino acid uptake rate to be limiting the growth rate of pelagic bacteria (thus ignoring inorganic nitrogen sources). Bulk amino acid concentrations in the upper ocean are on the order 10^{-9}–10^{-7} M (e.g., Mopper and Lindroth 1982, Poulet et al. 1991). Assume a typical concentration of 10^{-8} M. For a 0.5-μm radius bacterium, equation 2.21 predicts an uptake rate of ca. 5×10^{-15} mol d^{-1}. (Again, as an exercise, try to do the computation yourself. Assume $D = 10^{-5}$ cm^{2}s^{-1}). Because a bacterium contains on the order of 10^{-15} mol amino acids (and N), and if amino acids are the only source of N for the bacteria, this corresponds to a maximum growth rate of 5 d^{-1}. This is more than typical bacterial growth rates in the ocean. However, maximum measured amino acid uptake rates in field populations of bacteria are typically substantially less than the maximum estimated by equation 2.21 (e.g., Fuhrman and Ferguson 1986, Suttle et al. 1991). It has been suggested that the lower than expected uptake rate could be a result of diffusion impairment caused by a thick mucus sheet and a thick outer membrane found in many bacteria (Koch 1997), although for small molecules, a gel should not provide much impedance (Ploug and Passow 2007). However, bacteria are not necessarily perfect absorbers; hence, $C_a > 0$, and equation 2.21 thus provides an upper estimate of the uptake rate. Note that both of the above sample calculations have assumed nitrogen to be the limiting element, but they would work equally well with any limiting element.

In any case, cells that are, for example, ten times larger than bacteria (many heterotrophic flagellates) would have ten times higher diffusion-limited uptake rates. But such cells would have volumes $10^3 = 1000$ times larger and, therefore, a maximum diffusion-limited specific growth rate 100 times less. In the real world, heterotrophic flagellates have growth rates that are similar to that of bacteria, which implies that they cannot derive significant nutrition from dissolved organic material. Consequently, they feed on particles.

2.6 MAXIMUM AND OPTIMUM CELL SIZE

The metabolic expenditure of a unicellular organism scales approximately with its radius squared (at least with a power >1), whereas in

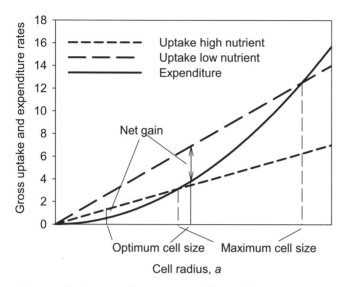

Fig. 2.6. Optimum and maximum cell size of an osmotroph. For diffusive supply, uptake rate increases linearly with cell radius, and the slope is proportional to the ambient nutrient concentration (dashed lines). Expenditure (metabolism) increases as a power function of cell size. The maximum possible cell size is the size at which uptake equals expenditure, i.e., where the uptake and expenditure lines intersect. Similarly, optimum cell size, where absolute growth (uptake minus expenditure) is the largest possible, is at the cell size where the difference between uptake and expenditure curves is the largest. Both optimum and maximum cell sizes increase with ambient nutrient concentration.

osmotrophs, solute uptake is proportional to cell radius (eq. 2.21). Because the growth (mass increase) of a cell equals the difference between its uptake and expenditure, there is a cell size at which the net gain is optimized and, similarly, a maximum possible cell size where gain and expenditure are equal and the cell cannot grow any further (Jumars 1993, fig. 2.6). When the maximum cell size has been reached, the cell can continue mass deposition only by dividing. Maximum net gain rate can be maintained if cell divisions are adjusted such that average cell size equals optimum cell size. Optimum and maximum cell sizes will increase with increasing ambient concentration of solutes (fig. 2.6). Jumars (1993) suggested that this may explain why bacteria in laboratory cultures grow very large; this is not a pathological effect but simply one of high nutrient availability. This could also be part of the explanation of why bacteria attached to particles are generally much larger than those free in the water (Alldredge et al. 1986): the attached bacteria are not necessarily different from those in the ambient water, but local nutrient availability may simply be much higher on the particle surface. In diatoms, daughter cells are always smaller than the mother cell, and cell size in a cell line therefore declines with the number of divisions. Diatom cell size can increase again only after sexual reproduction. During a phytoplankton bloom, dividing diatoms therefore continuously decrease in size; this happens concurrently with the exhaustion of the ambient concentration of inorganic nutrients, and Jumars (1993) speculated that optimal (or maximal) cell size might be tracking ambient nutrient concentration. This idea remains untested. However, the declining cell size will allow a population of diatoms to survive longer in the plankton than if cells had remained at the initial large size, and it will permit a large final population size at bloom termination. This, in turn, will produce a larger population of resting spores and, hence, improve conditions for recolonization of the water column once conditions again become favorable for diatom growth.

The above consideration might suggest that optimum cell size is not necessarily the smallest possible cell size, as argued in the preceding section. However, there is no inconsistency here. The optimum-size argument is valid only within a species, not among populations of species competing for a limiting resource. The cell size of a species may vary according to nutrient conditions as suggested, but when it comes to competition among species, it is the specific growth rate rather than the absolute net gain that matters. When nutrient uptake is diffusion limited, the specific uptake and growth rates decline with increasing size, as suggested by equation 2.24. The species that divide the fastest will win the competition for a common, limiting resource.

2.7 DIATOMS: LARGE YET SMALL

The above considerations show that large osmotrophs are more likely to become nutrient limited than small ones and that small size therefore can be considered an adaptation to low nutrient availability and to efficient competition for limiting nutrients. However, the diffusion limitation of nutrient uptake may be partly circumvented by organisms that inflate their size, for example, by having a big vacuole with a low concentration of limiting elements (e.g., nitrogen). Thus, doubling the cell radius without increasing the cell content of nitrogen reduces the volume-specific nitrogen content (C_N') by a factor of 8, but the volume-specific nitrogen uptake by diffusion (-Q_N') by only a factor of 4 (eq. 2.23), and, consequently, increases the growth rate by a factor of 2 (eq. 2.23). The advantage in practice will be less because the vacuole cannot be completely devoid of limiting elements, and therefore, the enhancement of mass-specific nutrient uptake and growth rate by this process at the very best increases in proportion to cell radius. Such a size inflation or "large yet small" strategy has been proposed for diatoms that have a large vacuole in the central part of the cell and with the cytoplasm organized as a 1 μm to several-micrometers-thick layer along the cell wall (Thingstad 1998). The strategy may be more generally adopted by phytoplankton cells, albeit in a less extreme form, as evidenced by the general decline in volume-specific mass content with increasing cell size found for phytoplankters (Menden-Deuer and Lessard, 2000). The inflated size brings an additional competitive advantage because size provides a partial refuge from predation (chapter 8). Normally, one assumes a trade-off between efficient predator defense (large size) and efficient nutrient uptake (small size) in unicellular osmotrophs, but the inflated size strategy combines both advantages and has therefore also been termed a "Winnie-the-Pooh" strategy (Thingstad et al. 2005) [". . . and when the Rabbit said, 'Honey or condensed milk with your bread?' he was so excited that he said, 'Both.' (Milne 1926, 37]. Inflated size strategies are found among other plankton organisms, such as salps and jellyfish, that expand their sizes by large gelatinous or mucous feeding structures. Although the mechanisms are different, they serve a similar purpose: to increase the rate of food acquisition and decrease the mortality rate from predation. Yet another example may be provided by the dinoflagellate *Noctiluca*, which has an extraordinarily low mass content per volume (near two orders of magnitude less other phytoplankters; Buskey 1995). In this species, positive buoyancy (Kesseler 1966) in combination with an inflated size allows for rapid ascent in the water and a consequently high encounter rate with prey (see also legend of fig. 4.2). Again, the inflated size allows for enhanced nutrient acquisition and, potentially, lower mortality.

Several colonial phytoplankters, such as the diatom *Chaetoceros socialis* and members of the often dominating cosmopolitan Prymnesiophycean genus *Phaeocystis*, have cells organized on the surface of a sphere that encompasses a large volume of cell-free water and thus is functionally similar to the way that diatoms have their cytoplasm organized. Can this similarly be considered an adaptation to nutrient uptake? Because cell number in a colony scales with colonial radius squared (surface area), whereas diffusion-limited nutrient uptake scales with its radius, the nutrient transport per cell varies inversely with colony radius (Thingstad et al. 2005). So, from the point of view of nutrient acquisition, colony formation is a disadvantage, although turbulence may partly compensate for this disadvantage (chapter 3).

There are also penalties associated with the inflated size strategy, particularly if large size is attained by the inclusion of a big vacuole (Raven

1997). It takes extra material and energy to build and maintain a vacuole. Also, a large vacuole may require a silicate wall for protection and support, which, in turn, imposes a need for silica and a higher sinking velocity on diatoms. There are, therefore, likely to be quite severe mechanical and other constraints to the size of the vacuole and, hence, limitations to the large-yet-small strategy. In diatoms, the volume-specific mass content is less than that of other phytoplankters, especially for large cells, but the difference even for the largest ones is less than a factor of 5 (Menden-Deuer and Lessard 2000), suggesting that an increase in diameter by a factor of $5^{1/3} = 1.7$ represents an upper limit to how far the size-inflation strategy can be taken.

2.8 DIFFUSION FEEDING

Several unicellular pelagic heterotrophs are nonmotile and thus must depend on the prey coming to them rather than vice versa. Heliozoans, helioflagellates (fig. 2.7), and foraminiferans, for example, feed on suspended bacteria. Because the cells are largely spherical, and because the motility of bacteria can be described as a diffusive process, prey encounter rates can be computed from equation 2.21. Consider as an example the helioflagellate *Ciliophrys marine* (fig. 2.7). It may be difficult to assign an exact encounter radius to this helioflagellate; obviously, it is larger than the cell radius (~4 µm) and smaller than the length of the sticky pseudopodia. If we conservatively take the cell radius as a measure of the dining sphere, and assume 10^6 bacteria ml^{-1} and diffusivity 10^{-5} cm^{2}s^{-1} (cf. above), then equation 2.21 predicts a prey encounter rate of 180 bacteria helioflagellate^{-1} h^{-1}. We can evaluate this number by estimating the magnitude of the growth rate that such an ingestion rate would support. To do this, assume that the flagellate has a radius ten times the radius of the prey and, therefore, a cell volume one thousand times that of the prey; assume further that the flagellate has to eat three times its own volume per cell division (this is a growth yield of 1/3); then the helioflagellate can perform a cell division every 16.5 h. This is a realistic growth rate, and this "back-of-the envelope" calculation thus demonstrates that ambush feeding in helioflagellates and similar protists is a feasible food acquisition strategy at typical oceanic bacterial concentrations. This feeding strategy has been termed diffusion feeding (Fenchel 1984).

Even nonmotile particles may be encountered by a diffusion feeder because small particles undergo thermally driven Brownian motion. For particles smaller than about 1 µm, Brownian motion leads to sizable diffusivities that can be estimated from

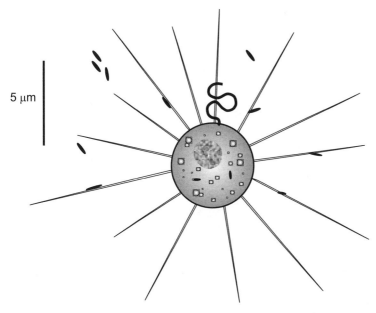

5 μm

Fig. 2.7. Helioflagellate (*Ciliophrys marina*=*C. infusiunum*), a diffusion feeder. The flagellum is inactive or moving very slowly in a figure-of-eight posture. The cell feeds on bacteria, which are caught on the tentacles, transported to the cell body, and ingested. Artwork by S. Jonasdottir.

$$D = KT/6\pi\eta a \qquad\qquad (2.26)$$

where K is Boltzmann's constant (1.38×10^{-16} g cm^{2}s^{-2} °K^{-1}), T is absolute temperature (°K), and η is the dynamic viscosity ($\sim 10^{-2}$ g cm^{-1} s^{-1}). In recent years it has been realized that small particles occur abundantly in the ocean (Wells and Goldberg 1991, Koike et al. 1990) and that they may be important as a food source for flagellates (e.g., Gonzáles and Suttle 1993, Tranvik et al. 1993). For example, colloids ($\sim 10^{-6}$ cm radius) occur at typical concentrations of 10^9 particles cm^{-3} with carbon concentrations of the same order as the carbon concentration of phytoplankton ($\sim 100\,\mu$g C L^{-1}; Wells and Goldberg 1991). The diffusivity of colloids is about 2×10^{-7} cm^2 s^{-1} according to equation 2.26, leading to significant diffusional encounter, and hence feeding, rates on such colloidal material for small flagellates. [As an exercise, compute the potential ingestion rate (in units of μg C d^{-1}) of colloids by a 4-μm radius diffusion feeder. Assume an ambient colloid concentration of $100\,\mu$g C L^{-1}. Evaluate the result relative to the carbon content of the cell, $\sim 2 \times 10^{-5}\,\mu$g C.]

Even in flagellates that generate a feeding current, diffusion may be a more important encounter mechanism than direct interception for colloidal-sized particles (Shimeta 1993). We shall return to the combined effect of diffusion and advection in the next chapter.

2.9 NON-STEADY-STATE DIFFUSION: FEEDING IN NAUPLII

From arguments identical to those for unicellular organisms feeding on dissolved organics, diffusion feeding is feasible only in small organisms. However, a variant of diffusion feeding is practiced by many copepod nauplii, which are much too large for "ordinary" diffusion feeding to be profitable. These nauplii do not have a feeding current, nor do they cruise through the water. Rather, they hang motionless in the water while collecting flagellates, and other organisms that are "diffusing" toward them. At short intervals, they jump to a new position. Apparently they move to a new position when the local food source has been depleted. Clearance rates computed from equation 2.22 are insufficient to account for the observed clearance rates. However, equations 2.20–2.22 are valid only at steady state. When a nauplius jumps to a new position, diffusional encounter is initially much higher than predicted by equation 2.21 because the prey concentration in the immediate environment of the nauplius is high and the concentration gradient much steeper than at steady state. Only as time passes does the concentration in the immediate surrounding of the nauplius become depleted, and the concentration gradient approaches that described by equation 2.19 at steady state. The time-dependent version of equation 2.22 is (Osborn 1996)

$$\beta_{\text{diffusion},t} = 4\pi Da\left[1+\left(\frac{a}{(\pi Dt)^{0.5}}\right)\right] \tag{2.27}$$

where t is the time, and the term in the squar bracket is the time-dependent part. The bracketed term goes toward 1 as $t \to \infty$, so for long t, equation 2.27 becomes similar to equation 2.22.

Equation 2.27 has more general application, and it is worthwhile to examine for instance, how much the flow is enhanced and how long it takes to reach a "semi"-steady state. Steady state is approached very quickly for small, micrometer-sized collectors, and the enhancement of the flow is insignificant, whereas steady state is approached only slowly for millimeter- to centimeter-sized ones, where the enhancement is also substantial (fig. 2.8A). For example, with a diffusivity of 10^{-5} cm^{2}s^{-1}, the flow comes within 10% of the steady-state flow in 0.03 s for a 1-μm radius collector, 30 s for a 0.1-mm one, but it takes 3×10^6 s, ~1 month, for a 1-cm collector. Thus, although steady-state considerations are appropriate for small organisms, they rarely are for larger ones.

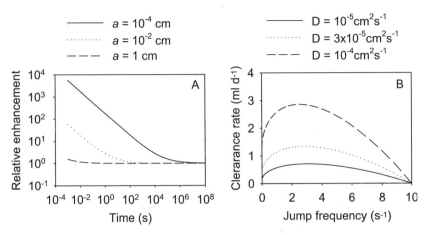

Fig. 2.8. Non-steady-state diffusion. A: Relative time-dependent enhancement of diffusive transport to a spherical collector over that at steady state as a function of time. The non-steady-state enhancement is insignificant for small collectors, where steady state is also approached rapidly, but it is substantial in large collectors, where steady state is approached only slowly. The computations have been made using equation 2.27 and assuming a diffusion coefficient of $D=10^{-5}$ cm^{2}s^{-1}. B: Estimated clearance rates of jumping copepod nauplii (*Acartia tonsa*) that perform "interrupted" diffusion feeding (eq. 2.28 assuming a duration of a jump to be 0.1 s and, hence, $\phi=0.1$ s/T). Various prey diffusivities have been assumed. The clearance rate depends on the jump frequency $(1/T)$ and has a maximum at about 3 jumps s^{-1}. Panel B modified from Titelman and Kiørboe (2003).

We can use equation 2.27 to analyze feeding in the jumping nauplius. If the nauplius jumps after a time interval T, and if it spends a fraction of its total time, ϕ, jumping (i.e., not feeding), then the time-averaged encounter-rate kernel is:

$$\beta_{average} = (1-\varphi)\frac{1}{T}\int_0^T \beta_t dt$$ (2.28)

$$= 4(1-\varphi)\pi Da\left(1+\frac{2a}{(\pi DT)^{0.5}}\right)$$

where β_t is given by equation 2.27. This function varies unimodally with $1/T$ (jump frequency) (fig. 2.8B). If the nauplius jumps too often, it spends all its time jumping rather than feeding. Conversely, if it jumps too seldom, prey-encounter rate becomes severely diffusion limited. There is, therefore, a jump frequency at which prey encounter rate peaks. For nauplii of the copepod *Acartia tonsa* (radius $a=0.01$ cm) feeding on motile flagellates (with diffusivities of order 10^{-4}cm^{2}s^{-1}), this optimum is about three jumps per second (fig. 2.8B). This predicted jump frequency is similar to the jump frequency realized by these nauplii, and the clearance estimated from equation 2.28 similar to that observed (Titelman and Kiørboe 2003). One may argue that this is a complicated way of saying something simple: move to a new place when the present place is depleted. But the use of a model allows a quantitative prediction and therefore a more detailed understanding of the feeding behavior of copepod nauplii. Also, this exercise allows us to introduce equation 2.27, which we use in the next example.

2.10 BACTERIA COLONIZING A SPHERE

We learned above that many pelagic bacteria are motile. One can wonder why this is so. Diffusion completes the job of bringing solutes to the cell, and in chapter 3 we shall see that swimming does not increase the rate at which solutes are transported to small cells. So why should bacteria spend energy swimming? One reason is that many pelagic bacteria attach to particles, including the large (millimeter to centimeter) particle aggregates known as "marine snow" (Lawrence et al. 1995, Fenchel 2001). By means of exoenzymes, the attached bacteria solubilize the component particles and feed on the resulting dissolved organic matter (Smith et al. 1992). To find and colonize a particle, the bacteria need to

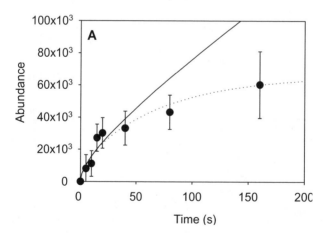

Fig. 2.9. Bacteria colonizing agar sphere. The initial colonization rate is well predicted by equation 2.27 (the full line). After a while, accumulation rate declines because bacteria also detach from the sphere. The dotted line is a fit to the data of a model that considers both colonization and detachment. The experiment used bacteria that had originally been isolated from suspended particles. Modified from Kiørboe et al. (2002).

swim. Because the motility of bacteria can be described as diffusion, we can use equations 2.27 and 2.21–2.22 to predict initial and steady-state colonization rates, and we can compare the prediction with observations. Figure 2.9 gives an example of the rate at which motile bacteria accumulate on the surface of a 0.2-cm radius sphere. Initially, the accumulation rate is well predicted by equation 2.27 (later, accumulation rate vanishes because the bacteria start to jump off the particle again, but that is a different story). The main point here (in addition to illustrating eq. 2.27) is that bacteria accumulate very rapidly on the surface of suspended particles and that this accounts for the several orders of magnitude higher concentration of bacteria found in marine snow than in the surrounding water (Alldredge and Silver 1988). Marine snow aggregates are microbial hot spots. We shall return to the story of particle-attached bacteria and microbial communities in chapters (3, 4, and 7).

2.11 EFFECT OF SHAPE

In all of the above it has been assumed that all organisms and particles are spherical. Although many phytoplankters and bacteria are near-spherical, and the assumption thus well approximated, rod-shaped bacteria and elongate or particularly chain-forming algae are common and deviate significantly from spherical shape. Does that change the above considerations significantly? Berg (1993) showed that the diffusive flow to a cigar-shaped (ellipsoid) adsorber with semiaxes a and b, and with $a^2 >> b^2$, is

$$Q = 4\pi DaC_\infty / \ln(2a/b) \qquad (2.29)$$

That is, the flow is smaller than that for a sphere by a factor $\ln(2a/b)$. As pointed out by Berg (1993), this factor is not very large; even for a very elongate, thin object, 10^4 times longer than thick ($a = 10^4 b$), it is less than 10. This implies that the solute flow becomes roughly proportional to the length of the cell or chain, not to its surface area. That is, similar to the situation for a sphere.

The question of interest is whether—and how much—the specific diffusion-limited uptake rate changes as a function of change in shape. If one accepts the ellipsoid as a fair approximation of the shape of elongate cells and even cell chains, then we can compute the volume-specific diffusion-limited uptake rate. The shape can be quantified as the aspect ratio of the ellipsoid, i.e., the ratio of its longest to its shortest linear dimension, a/b. The volume of an ellipsoid is $\frac{1}{3}\pi ab^2$. It turns out that the volume-specific diffusion-limited flow is larger for an ellipsoid than for a sphere of the same volume, but the effect is rather modest. For example,

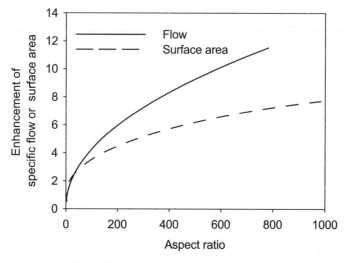

Fig. 2.10. Effect of shape. Enhancement of diffusion-limited volume-specific solute flow and surface area in a prolate ellipsoid relative to that of a sphere of the same volume, both plotted as a function of the aspect ratio (ratio of longest to shortest dimension).

an aspect ratio of 10 enhances the flow by a factor of ~1.4 and by a factor of 13 for an aspect ratio of 1000 (fig. 2.10). However, it is high enough that one may consider elongated shape as an adaptation to competition for limited resources, and there is evidence that some bacteria adapt to (modest) starvation by elongating (Steinberger et al. 2002) (many bacteria adapt to severe starvation by shrinking and closing down their cellular machinery; see section 6.9).

Cell stretching is not without cost. It is well known that a sphere has the lowest possible surface area for a given volume, and stretching thus increases the specific surface area of the cell. For an ellipsoid, the surface area per unit volume increases with the aspect ratio almost as much as the specific solute flow. Because there are expenses associated with formation of cell walls, the advantage of stretching thus appears to be rather limited, and nonspherical shapes may be related to the optimization of other cell functions, such as the capability of directional swimming in small cells (Dusenbery 1996).

2.12 FLUX FROM A SPHERE (OR A POINT SOURCE): CHEMICAL SIGNALS

In all our considerations so far, we have considered diffusion of substances toward an absorbing sphere. But all the equations describe equally well

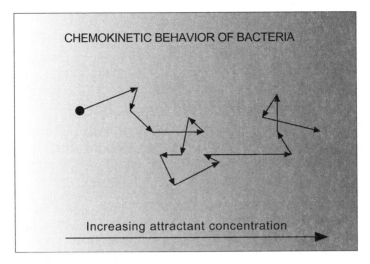

Fig. 2.11. Chemotactic behavior in motile bacteria, which is possible because of their run-tumble swimming mode. There will be a net movement toward regions of highest attractant concentration if the length of a run increases with increasing attractant concentration during a run. This allows the bacteria to accumulate in regions of high concentrations of attractant molecules, e.g., dissolved organic matter. Chemotactic behavior is possible only in relatively large bacteria because bacteria smaller than ~0.6 µm diameter lose directional persistence during runs as a result of Brownian rotation (Dusenbery 1996). Similarly, rodlike shapes allows for more precise steering, which may explain why rodlike shapes and this mechanism of chemotactic behavior are common among pelagic bacteria (Dusenbery 1998).

diffusion of substances away from a leaking sphere. The leaking substance could be dissolved organics leaking from a marine snow aggregate. Because many bacteria have chemotactic behavior (fig. 2.11), allowing them to swim up a concentration gradient of attractor molecules, such leakage would provide a signal for colonizing bacteria, and there are several demonstrations that pelagic bacteria in fact do accumulate around small-scale nutrient sources and leaking particles (Blackburn et al. 1998, Kiørboe et al. 2002, Barbara and Mitchell 2003a). Or it could be an egg leaking signal molecules that would guide chemosensory sperm cells toward the egg, as has been demonstrated in many marine invertebrates (e.g., Ward et al. 1985) and in humans (Spehr et al. 2003). In general, chemical signals in the ocean are important in many instances to help improve encounter rates, be it between predator and prey or between mates. In these cases, we would be interested in the spatial distribution of solutes around the sphere (as usual, we consider our organisms of interest to be spherical). The steady-state spatial distribution of solutes diffusing from a leaking sphere can be described from equation 2.19, which is:

$$C_r = \frac{Q}{4\pi Dr} + C_\infty \quad \text{for } r \geq a$$

(2.30)

Note that this equation applies to both absorbing and leaking spheres (illustrated in fig. 3.3A); for an absorbing sphere, Q is negative, and the concentration increases away from the sphere; for a leaking sphere, Q is positive (=leakage rate), and concentration decreases away from the sphere. In the case of leaking pheromones or other "special" molecules, the background concentration, C_∞, is typically zero. Note also that the concentration distribution is independent of the size of the sphere. Thus, equation 2.30 also applies to continuous point sources (or sinks), i.e., for $a = 0$.

Equation 2.30 is relevant for examining chemical signals only in the case of very small-scale phenomena and small organisms because it applies only to steady state. We saw above that the times required to reach "near"-steady-state solute distributions for millimeter and larger particles are on the order of hours to days. Copepods, for example, would have moved long distances in such time intervals, making diffusion insignificant in distributing the chemicals. A swimming copepod leaking pheromones, for example, would rather paint a chemical trail in its wake. To analyze this and many other situations, we need to consider advection in addition to diffusion as a means of mass transport. This is the topic of the next chapter.

Chapter Three

DIFFUSION AND ADVECTION

3.1 MOVING FLUIDS

UNTIL NOW we have considered only the diffusive transport of matter—or transport of organisms with a motility that can be described by a diffusion formalism. However, water moves, and water motion may advect matter. In regard to transport to and from planktonic particles or organisms, flow per se will not cause any transport because the plankters are embedded in the general flow. Advective transport occurs only when there is a velocity difference between the particle (or organism) and the ambient water. Such a velocity difference occurs when the particle moves (swims, or sinks or rises as the result of a density difference), or when there are velocity gradients in the ambient water, e.g., from turbulence. Advection may enhance mass transport and, therefore, for example, nutrient uptake rates and spreading of chemical signals. In this chapter we consider the combined effects of diffusion and advection for encounter rates (in a general sense) in situations where diffusion is still important. In chapter 4 we examine situations where advection dominates and diffusive processes can be disregarded.

3.2 VISCOSITY, DIFFUSIVITY, Re, AND Pe

The environment at the spatial scale of most plankters is dominated by viscous forces, whereas inertial forces can be largely ignored. This implies that water is "sticky" and thick as syrup. Thus, if you try to reach out after a particle with an appendage, the surrounding water and the particle will just move away as you approach it; you withdraw the appendage and the particle moves back again. The flow is reversible. This of course poses some problems for an animal that feeds on particles. But it also implies that any motion you make causes motion in the ambient water—you draw water with you, and you generate velocity gradients in the ambient water. As we shall see in chapter 5, these velocity gradients may be perceived by others—friend and enemy. High viscosity and lack of inertia also imply that as soon as you stop propelling yourself, you stop. And as soon as you stop, motion of the ambient water stops. In other words, the fluid disturbance (or, in some contexts, the signal) you generate propagates fast and dissipates fast. The diffusivity of momentum is called kinematic viscosity and is characterized by a viscosity coefficient, v, that has the same dimensions as molecular diffusivity (L^2T^{-1}). The relative significance of inertial to viscous forces is characterized by the dimensionless Reynolds number

$$Re = ua/v \tag{3.1}$$

where a is a linear dimension of an object moving at velocity u[1]. For $Re < 1$ viscous forces dominate, and inertia can be largely ignored. The kinematic viscosity, v, of seawater is approximately 10^{-2} cm^2s^{-1}, and we can thus compute Re for various plankters. For example, for a 1-μm bacterium swimming at $50\,\mu m\ s^{-1}$, $Re = 5 \times 10^{-5}$, and for a 30-μm ciliate swimming at $50\,\mu m\ s^{-1}$, $Re \approx 10^{-3}$, in both cases well below 1. For a 1-mm copepod generating a feeding current with a velocity of $1\,cm\ s^{-1}$, $Re = 10$, and for a 0.5-cm marine snow aggregate sinking at $0.1\,cm\ s^{-1}$, $Re = 5$. In these latter cases, we cannot really ignore inertia. (We will sometimes do it anyway, at least in the first place, and later show that in some contexts the error we make is small.) The point is that small plankters operate at small Re well below 1, whereas marine snow, mesozooplankton, fish larvae, and all those plankters that we can see with the naked eye operate at smallish Re, near or a little above 1. In all cases the generated flow is laminar, i.e., well ordered and nonturbulent. Turbulent flow is generated only at very much higher Reynolds numbers. Recall that flow in a pipe becomes turbulent at $Re \approx 2000$.

Solutes (and chemical signals) behave similarly yet differently from hydromechanical disturbances (signals). Chemicals can spread by molecular diffusion and by advection. We have already seen that the diffusivities of most biological molecules (nutrient salts, oxygen, amino acids, sugars) are on the order of 10^{-5} cm^2s^{-1}. This is three orders of magnitude less than the diffusivity of momentum (kinematic viscosity) in seawater. Hence, chemical signals spread slowly by diffusion but also dissipate slowly. They are persistent, at least on spatial scales exceeding 1 mm. Recall that we can estimate a time scale for diffusion from dimensional analysis: $D \approx L^2T^{-1} \Rightarrow T \approx L^2D^{-1}$. For example, the time scale for diffusion at a spatial scale of 1 μm is 10^{-3} s, whereas the time scale for diffusion at a spatial scale of 1 mm is 1000 s (1/4 h), and that at 1 cm is 1 d. Thus, diffu-

---

[1]Warning: In calculating Re and Pe for (spherical) objects, one tradition uses radius, and one uses diameter. That can cause confusion. Following Karp-Boss et al. (1996), I consistently use radius.

sion is very rapid at very small scales but becomes extremely slow at, for example, copepod spatial scales. Evidently, diffusing chemical signals would not be very useful in detecting and locating an approaching predator! Although chemical signals may be evidence of the general presence of predators, we would predict that hydrodynamic or visual signals are required to locate an approaching predator.

But solutes may also spread by advection, and here again we have a dimensionless number to characterize the relative significance of advective to diffusive transport, the Péclet number Pe:

$$Pe = ua/D \qquad\qquad (3.2)$$

Pe is a quantitative indicator of the relative importance of advection versus diffusion in moving solutes a distance a through the fluid.[2] Thus, for $Pe < 1$, diffusion dominates, whereas for $Pe > 1$, advection begins being important. For example, for the 1-μm bacterium swimming at 50 μm s^{-1} feeding on solutes with a diffusivity of 10^{-5} cm^{2}s^{-1}, $Pe = 5 \times 10^{-2}$. Thus, diffusion dominates, and swimming, therefore, probably will not facilitate nutrient uptake in bacteria. In most of the cases of relevance for chemical signaling that come to mind, $Pe \gg 1$. For example, for a 1-mm (radius) copepod swimming at 5 mm s^{-1} while exuding pheromones with a diffusivity of 10^{-5} cm^{2}s^{-1}, $Pe \approx 5000$. This implies that the chemicals exuded by the swimming copepod may paint a trail: the trail is long and spreads only slowly by diffusion. To visualize, think of a jet plane "exuding" water particles. This is a high-Péclet-number situation because the plane moves fast while the water particles diffuse slowly, and as a result, the trail of the jet can be seen very far behind the plane and for a long time. We shall be more explicit and precise about these and other examples below.

3.3 FLOW AROUND A SINKING SPHERE

Before we can examine how advection may influence the transport of diffusing substances, we need to describe the fluid flow around our moving plankter. This description will also prove useful for other applications later. To describe the flow one needs to solve the so-called Navier–Stokes equation. This equation follows directly from Newton's second law and describes how velocities in a fluid vary spatially and temporally as a

---

[2]Strictly speaking, it is incorrect to apply Pe to a particle because Pe is a property of the fluid, not of the partticle. The best one can do is to apply Pe to a parcel of water of the same size as the particle.

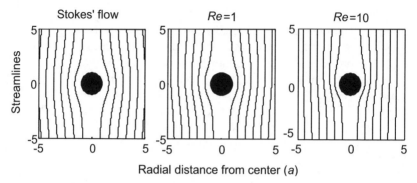

Fig. 3.1. Streamlines around a sinking sphere at $Re \approx 0$ (Stokes' flow), $Re=1$, and $Re=10$. Distances are in units of particle radii. With increasing Reynolds number the flow becomes increasingly asymmetrical. Modified from Kiørboe et al. (2001).

function of forces (viscous forces, gravitational forces, and pressure forces). It can therefore be used to describe the flow around a moving object of any size or shape. Only in a few simple cases, however, can the equations be solved analytically (or approximated by an analytical solution).[3] One of these cases is that of a sinking sphere at low Re. Fortunately, most plankters operate at low Re and are sufficiently spherical to make such a simplification meaningful. We shall later provide the actual equations; here we consider only the general pattern for laminar flow at very low and small Re.

The streamlines around a sinking sphere at low ($Re \ll 1$; Stokes' flow) and smallish Re (1 and 10) are illustrated in figure 3.1, and the fluid velocities (relative to the sphere) in figure 3.2. The flows for $Re > 0$ were found by solving the Navier–Stokes equation numerically. For Stokes' flow, the flow is very regular and exactly symmetrical. Note that the velocity of the flow at the surface of the sphere in all cases is 0. This is the "no-slip" condition. As we go away from the sphere, either upstream or downstream or to the sides, the flow velocity increases and approaches the ambient flow velocity (= the sinking velocity of the sphere). However, even six particle radii away, the flow velocity is only about 80 % of the velocity far away; that is, the sinking particle affects the flow, or disturbs the water, at a long distance. With increasing Re, the flow becomes asym-

---

[3]An *analytical* solution means that the solution can be written as an equation. Many equations do not have analytical solutions and must, thus, be solved *numerically* for each specific case.

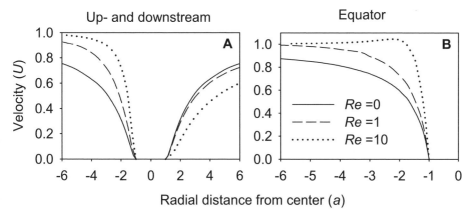

Fig. 3.2. Flow velocities around a sinking sphere as a function of the distance to the sphere center in the up- and downstream directions and off the side (equator) of the sphere. Flow velocities are relative to the particle and expressed as a fraction of the sinking velocity of the sphere. From Kiørboe et al. (2001).

metric (fig. 3.1), the velocity gradients become steeper, and the sinking particle affects ambient flow only to shorter distances (fig. 3.2).

3.4 MASS TRANSPORT TO A SINKING SPHERE

We are now ready to examine the mass transport by the combined effects of diffusion and advection to a sinking spherical collector (we treated the stationary sphere in chapter 2). Recall that there will be a region around the stationary sphere where solutes are depleted because they are taken up by the sphere. Because of advection, the solute-depleted region surrounding the particle will constantly be replaced with new water, and the higher the flow, the faster the replacement and the higher the transport. The dimensionless Péclet number indicates the relative significance of advective over diffusive transport. But because it applies to the fluid, not to the particle, this number cannot directly be used to quantify the solute transport to a particle in an advective situation—it can only be considered indicative of the significance of advection. What we really want to know is how much or by what factor the solute flow to a particle is increased over diffusive transport in a particular situation. What advection really does is that it increases the steepness of the concentration gradient in the boundary layer around the particle. Because fluid flow at the very surface of the sphere is zero (the "no slip" condition), transport at the surface is only by diffusion. And from Fick's first law we know that the flux is proportional to the concentration gradient. Thus, the solute flow is exactly proportional to the concentration gradient at the very cell surface: if we double the concentration gradient, then we double the solute flow. The factor by which the average concentration gradient at the cell surface is increased as a result of advection is termed the Sherwood number (yet another of these many dimensionless numbers). The Sh is thus the ratio of total to diffusive flux. Please note that the (global) Sherwood number applies only if the solute flow is transport limited. In flowing water (or for a sinking particle), equations 2.21 and 2.22 are modified to:

$$Q = -4\pi a D Sh C_\infty \tag{3.3}$$

and

$$\beta_{advection-diffusion} = 4\pi a D Sh \tag{3.4}$$

The Sherwood number has considerable practical interest and applicability in oceanography because, if we know it, we can evaluate the effect of advection on, e.g., nutrient uptake. The problem is to know it. Unlike the Péclet number, there is no easy way to compute it.

To estimate Sh we need to describe the concentration field around the sinking particle. If we know the concentration field, then we also know the concentration gradient at the surface of the particle, and then we know the solute flow and the Sherwood number. To describe the concentration field, we first describe the flow field (see above) and subsequently solve the advection-diffusion equation numerically (we shall not go into detail here). For low-Re situations, we can use Stokes' flow; for higher-Re cases (as applies, for example, to sinking marine snow), we need to solve the Navier–Stokes equations. Computed spatial distributions around a sinking sphere of substances with a diffusivity of $10^{-5}\,cm^2s^{-1}$ are shown in Fig. 3.3 for various values of Pe.

By numerically computing solute distributions and, hence, Sh for many different combinations of Re and Pe, one can generate approximate expressions for Sh. Here we provide expressions valid for $Re<0.1$ (Clift et al. 1978):

$$Sh = 0.5[1 + (1 + 2Pe)^{1/3}] \tag{3.5}$$

and for $Re>0.1$ and $30 < Pe < 50,000$ (Kiørboe et al. 2001):

$$Sh = 1 + 0.619 Re^{0.412} \left(\frac{v}{D}\right)^{1/3} \tag{3.6}$$

3.5 EXAMPLE: OXYGEN DISTRIBUTION AROUND A SINKING SPHERE

Before we apply the insights achieved above, let us just illustrate that the solute distributions predicted (fig. 3.3) in fact may describe real phenomena.

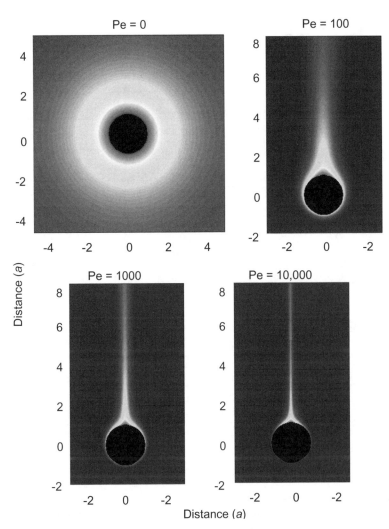

Fig. 3.3. Computed solute distributions around a sinking sphere at different Péclet numbers $(D=10^{-5}\ \mathrm{cm^2 s^{-1}})$. The equations used to compute the solute distributions—and the patterns shown—correspond both to spheres that absorb solutes (for example oxygen or inorganic nutrients) and to spheres that leak solutes (e.g., particles leaking dissolved organic matter). In both cases, dark corresponds to the concentration far from the sphere, which in the former case is high relative to the concentration near the sphere, and vice versa in the latter case. Modified from Kiørboe et al. (2001).

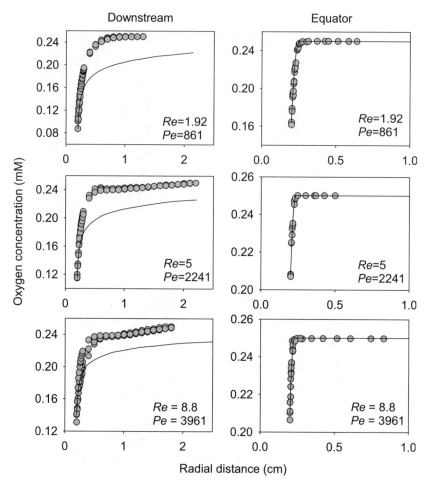

Fig. 3.4. Observed and predicted concentration fields of oxygen around a 4-mm-diameter sphere suspended in a flow field at various *Pe*. Predicted (lines) and measured (dots) concentrations of oxygen downstream and perpendicular to the flow off the equator of an oxygen-consuming sphere in laminar flow. Different flow velocities, corresponding to different sinking velocities, result in different *Re* and *Pe*. From Kiørboe et al. (2001).

In an experimental study, Kiørboe et al. (2001) used agar spheres suspended in a flow as models of sinking marine snow aggregates. The spheres were filled with yeast cells that consumed oxygen. There was therefore a constant flux of oxygen toward the sphere, and the distribution of oxygen can be described by solutions to the advection–diffusion equations, as illustrated in figure 3.3. The concentration of oxygen can be measured by microelectrodes along transects downstream of the sphere or perpendicular to the fluid flow off the equator of the sphere. Measurements and predictions correspond well with one another, particularly near the sphere and off the equator of the sphere (fig. 3.4). Downstream of the sphere, the correspondence is less convincing; this may simply result from a very narrow tail (fig. 3.3) that is meandering slightly in the flow, making it difficult to make measurements in the exact center of the invisible downstream tail.

3.6 Examples: Osmotrophs, Diffusion Feeders, and Bacterial Colonization of Sinking Particles

We can now use equations 3.5 and 3.6 to compute more precisely how much advection enhances transport to osmotrophs utilizing dissolved organics.[4] Take, for example, the 1-μm bacterium swimming at $50\,\mu$m s^{-1} and feeding on solutes with a diffusivity of 10^{-5} cm^{2}s^{-1} that we considered above. Its Pe was 5×10^{-2}, and its Sh is 1.016. That is, swimming enhances nutrient uptake by only 1.6%, i.e., insignificantly. The same holds true for sinking phytoplankton—in only the largest and most rapidly sinking cells does sinking significantly enhance nutrient acquisition (table 3.1). However, for organisms that depend on taking up diffusing molecules, the advantage of large size in terms of benefiting from an advective contribution to the transport of molecules is small relative to the disadvantage of large size in constraining the specific diffusive delivery of molecules. Generally, sinking and swimming play modest roles for nutrient uptake in small, unicellular organisms, as has also been realized previously (e.g., Purcell 1977, Sommer 1988, Karp-Boss et al. 1996).

The situation is different in larger particles, such as sinking marine snow (table 3.1) and colonial phytoplankters, such as *Phaeocystis* (Ploug et al. 1999). Here advection may increase the availability of nutrients or oxygen severalfold. This has practical implications. For example, if one wants to measure solute exchange between a marine snow aggregate and the ambient water, such as oxygen uptake rate (as a measure of microbial

---

[4]The Sherwood number computed for a sinking sphere does not strictly apply to a swimming organism because the flow fields around a sinking and a self-propelled sphere are different. The streamlines for a self-propelled sphere will come closer to the sphere surface than those for a sinking sphere; the concentration gradients are therefore steeper, and the Sherwood number higher. How much higher has not yet been addressed in the literature for realistic models of swimming microorganisms, only for a "squirmer," a highly artificial self-propelled sphere that severely overestimates the effect (Magar et al. 2003, Magar and Pedley 2005). However, the effect, even for the squirmer, is modest for small Péclet numbers, and we shall therefore ignore it here.

TABLE 3.1
Re, *Pe*, and *Sh* for Sinking Phytoplankton Cells and Marine Snow Aggregates

	Phytoplankton			
Radius, μm	Sinking velocity,[1] cm s^{-1}	Re (= au/v)	Pe (= au/D)	Sh[2]
0.5	2.3×10^{-5}	1.1×10^{-7}	1.1×10^{-4}	1.00
5	3.4×10^{-4}	1.7×10^{-5}	1.7×10^{-2}	1.01
50	5.0×10^{-3}	2.5×10^{-3}	2.5×10^{0}	1.41
500	7.5×10^{-2}	3.8×10^{-1}	3.8×10^{2}	5.06
	Marine snow aggregates			
Radius, mm	Sinking velocity,[3] cm s^{-1}	Re (= au/v)	Pe (= au/D)	Sh[4]
0.1	0.039	0.039	39	2.6
1	0.071	0.71	710	6.4
10	0.13	13	1300	18.8

[1] u (cm s^{-1}) = 2.48.a (cm)$^{1.17}$. Calculated from Stokes' law taking the declining cell density with cell size into account (Jackson 1989).

[2] *Sh* calculated using eq. 3.5 assuming $D = 10^{-5}$cm^{2}s^{-1}

[3] u (cm s^{-1}) = 0.13a (cm)$^{0.26}$ (Alldredge and Gotschalk 1988).

[4] *Sh* calculated using eq. 3.6, assuming v/D equal to 1000.

activity), it is important that one creates the correct fluid dynamic environment. That is, the aggregate or the colony needs to be suspended in a flow rather than resting in the bottom of a test tube. Otherwise, the measurements can be very misleading (e.g., Ploug and Grossart 1999).

Above we considered "diffusion" feeders, that is, nonmotile protozoans preying on bacteria or small particles diffusing toward them. However, even in flagellates that swim or generate a feeding current, for example, diffusion may be important in providing prey (Shimeta 1993) because of the combined effect of diffusion and advection. To see this, consider as an example a 5-μm diameter flagellate generating a 200 μm s^{-1} feeding current and feeding on colloidal-sized particles ($D \approx 2 \times 10^{-7}$ cm^{2}s^{-1}, eq. 2.26, section 2.7). The Reynolds number is 5×10^{-4}, the Péclet number is 25, and equation 3.5 then applies ($Sh = 2.4$). Equation 3.4 then predicts a volumetric encounter rate (β=clearance rate) of about 10^5 times its own body volume per hour, a value typical for clearance rates actually measured for nanoflagellates (e.g., Hansen et al. 1997). (Exercise: Try to make a similar computation with nonmotile, 0.5-μm bacteria as food, using the same equations.) Feeding in

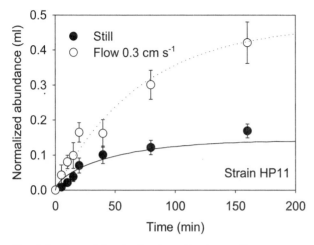

Fig. 3.5. Bacterial colonization of a 4-mm-diameter agar sphere in still and flowing water (0.3 cm s^{-1}). The enhancement of the initial colonization rate is well predicted by the Sherwood number as computed from equation 3.5. The abundance of attached bacteria is expressed relative to ambient concentration (normalized abundance). Note that in both cases, net colonization levels off as the rate at which cells detach comes into equilibrium with colonization rate. Modified from Kiørboe et al. (2002).

many flagellates is often considered to be mainly by direct interception, a process—as we shall see in the next chapter—that is not very efficient (Be ahead of the class: Try to calculate expected clearance rates by direct interception for the 5-μm flagellate using equation 4.3 provided in chapter 4.) These considerations, together with your own calculations, suggest that diffusion (in combination with advection) may be important in delivering small prey particles even to flagellates with strong feeding currents.

Finally, the Sherwood number concept also applies to organisms with random-walk motility. Thus, bacteria may colonize sinking marine snow aggregates faster than stationary aggregates, and the enhancement is well predicted by equation 3.5 (fig. 3.5).

3.7 EFFECT OF TURBULENCE ON MASS TRANSPORT: *Re, Pe,* AND *Sh* FOR TURBULENCE

Ambient water motion—turbulence—may also enhance mass transport to a solid body and, thus, for example, increase nutrient availability to (large) cells. The Sherwood number for turbulence can be computed in much the same way as we saw above for a sinking sphere by solving the Navier–Stokes and advection-diffusion equations numerically. We give only a very brief account here, based mainly on Karp-Boss et al. (1996), where more detail can be found. Turbulence first appears as large eddies that are strained and divided into subsequent smaller eddies. Because of the effect of viscosity, there will be a minimum size of such eddies (characterized by the so-called Kolmogorov scale) below which water motion is characterized by laminar shear (see section 4.6 for a more detailed description of turbulence). The minimum size of eddies is on the order of 1 mm or more for typical turbulent intensities in the ocean, and most unicellular and even colonial organisms thus experience turbulence as laminar shear, i.e., well-ordered or smooth velocity gradients. Because a solid body in a velocity gradient cannot travel with the same velocity as the ambient fluid, it will experience a velocity difference in much the same way as a swimming or sinking plankter. Shear has dimensions of T^{-1} (velocity difference per distance, $LT^{-1}/L=T^{-1}$), and the characteristic velocity difference experienced by a sphere in a sheared flow is thus the shear times the radius of the sphere:

$$u_{\text{shear}} = \gamma a \tag{3.7}$$

where γ is the shear rate. We can then define Reynolds and Péclet numbers for shear that are consistent with our previous definitions, i.e.,

$$Re_{\text{shear}} = \frac{u_{\text{shear}}a}{\nu} = \frac{\gamma a^2}{\nu} \tag{3.8}$$

and

$$Pe_{\text{shear}} = \frac{u_{\text{shear}}a}{D} = \frac{\gamma a^2}{D} \tag{3.9}$$

Karp-Boss et al. (1996) cites the following numerical approximations for the Sherwood number for turbulent shear (sphere):

$$Re_{\text{shear}} \ll 1,\ Pe_{\text{shear}} \ll 1:\ Sh_{\text{shear}} = 1 + 0.29\,Pe_{\text{shear}}^{0.5} \tag{3.10}$$

$$Re_{\text{shear}} < 1,\ Pe_{\text{shear}} \gg 1:\ Sh_{\text{shear}} = 0.55\,Pe_{\text{she}}^{1/3} \tag{3.11}$$

These equations, unfortunately, are not valid for intermediate values of Pe, and in this region one has to interpolate (Karp-Boss et al. 1996). Finally, the intensity of turbulence is often quantified as the rate at which turbulent kinetic energy dissipates as heat, the energy dissipation rate ε. The energy dissipation rate has dimensions of L^2T^{-3}, and shear and energy dissipation rate are related by $\gamma = (\varepsilon/\nu)^{0.5}$. Typical upper-ocean turbulent dissipation rates are up to 10^{-1} cm^{2}s^{-3} for a wind speed of 15–20 m s^{-1} or in areas with strong tidal mixing and much less in calm weather.

TABLE 3.2
Re, Pe, and *Sh* for Particles in Turbulence

Radius, μm	Re_{shear} $(=\gamma a^2/\nu)$	Pe_{shear} $(=\gamma a^2/D)$	Sh_{shear}
0.5	2.5×10^{-7}	2.5×10^{-4}	1.00
5	2.5×10^{-5}	2.5×10^{-2}	1.05
50	2.5×10^{-3}	2.5×10^{-0}	1.10
500	2.5×10^{-1}	2.5×10^{2}	3.5
5000	(2.5×10^{1})	(2.5×10^{4})	(16.1)

Reynolds, Péclet, and Sherwood numbers for turbulent shear for spherical particles of Varying size. A viscosity of $\nu = 10^{-2}$ cm^{2}s^{-1}, a diffusivity of $D = 10^{-5}$ cm^{2}s^{-1}, and a turbulent energy dissipation rate $\varepsilon = 10^{-2}$ cm^{2}s^{-3} have been assumed. The latter dissipation rate corresponds to the energy dissipation in the upper mixed layer at a wind velocity of 10–15 m s^{-1}. In computing Sherwood numbers for $Pe < 1$, eq. 3.10 has been used, and for $Pe > 10$, eq. 3.11. For $Pe = 2.5$, an interpolated result from the two equations has been used. Note that the equations do not apply at $Re > 1$, and, hence, figures computed for the largest particle (5000 μm) are very approximate (and reported in parentheses).

So equipped, we can now examine the significance of turbulence in enhancing mass transfer, etc., in spherical collectors, as above (table 3.2). Even for a relatively high turbulent energy dissipation rate of 10^{-2} cm^{2}s^{-3}, the enhancement is insignificant for phytoplankton-sized particles (Sherwood numbers near 1) and is significant only for large or very large (marine snow) particles when solutes and particles (motile bacteria) have a typical diffusivity of 10^{-5} cm^{2}s^{-1}. Thus, the conclusion resembles that drawn for sinking and swimming.

Small-scale turbulence may, however, enhance prey encounter rates in small "diffusion feeders" that prey on colloids and nonmotile bacteria with very low diffusivities. This was demonstrated experimentally by Shimeta et al. (1995). They measured clearance rates on nonmotile bacteria in the helioflagellate *Ciliophrys marina* that we considered in the previous chapter (fig. 2.7). Stillwater clearance rates were ~10^{-7} ml h^{-1}, but the clearance rate at a shear rate of 1 s^{-1} was about three times higher. Both the magnitude of the stillwater clearance rate and the enhancement in turbulence are well predicted by the equations provided here (eqs. 2.22, 3.4, and 3.11). (As an exercise: Make the computation yourself assuming bacterial size of 1 μm diameter and Brownian diffusion of the nonmotile bacteria.)

In the real world, plankters sink or swim and experience ambient turbulence at the same time, and the effects of both flows must be considered. In practice, either turbulence or sinking/swimming dominates the advective component of the transport (Karp-Boss et al. 1996), and as a first approximation, one need consider only the largest of these contributions.

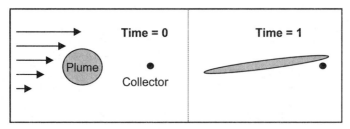

Fig. 3.6. A solute plume and a collector in a velocity gradient. A spherical solute plume (gray) is released in a velocity gradient at some distance from a suspended collector (black dot) at time=0. At time=1, the solute plume has been drawn out. The distance between the center of the plume and the collector is unaltered because both travel with the flow, but high concentrations have now come close to the collector. The resulting steep concentration gradient close to the collector makes diffusion efficient in transporting solutes to the collector. The flux toward the collector is consequently much higher at time=1 than at time=0.

The discussion above has considered the effect of turbulence on mass transport via its effect on renewing the water in the immediate vicinity of the collector and thereby increasing the concentration gradient at the very surface. Turbulence may, however, also affect mass transport by changing the distribution of particles/molecules that are unevenly distributed in the first place. Solutes and particles are often heterogeneously distributed in the ocean, even on small and intermediate scales. Solute plumes may exist around a leaking particle (fig. 3.3 and below), an excreting copepod, or a lysing cell. Similarly, organisms may release batches of sexual products in the water, and these batches then disperse and drift with the ambient flow. In a turbulent or sheared flow field, such plumes of solutes or particles will be stretched, and this may reduce the spatial distance between the collector and the plume or between the eggs and the cloud of sperm (fig. 3.6). At this smaller spatial distance, diffusive processes may do the rest of the job in delivering material to the collector or allowing encounters between sperm and eggs. On scales larger than the Kolmogorov length scale, eddies (vortices) may similarly stretch plumes into elongate filaments. Along this line of reasoning, Crimaldi et al. (2006) elegantly examined the effect of such vortex stirring for particle contact rate, with a view on benthic broadcast spawners that release both eggs and sperm freely in the water, and found a substantial effect, potentially accounting for the often surprisingly high fertilization rates of batch-spawned invertebrate eggs. We shall refrain from a more detailed analysis here and just note that this is another way that turbulent advection may enhance diffusive transports.

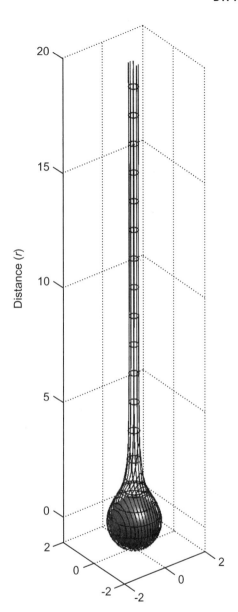

Fig. 3.7. Amino acid plume traveling behind a falling aggregate. Three-dimensional illustration of the plume deposited behind a 0.5-cm-radius aggregate. The depicted plume corresponds to the volume where amino acid concentration has been elevated by 3×10^{-8} M above ambient concentration. All distances are in units of particle radii. The plume is about 1 m long! Beyond 1 m behind the particle, diffusion has reduced amino acid concentration below the chosen threshold concentration. Modified from Kiørboe et al. (2001). Artwork by G. A. Jackson.

3.8 MARINE SNOW SOLUTE PLUMES: SMALL-SCALE HETEROGENEITY

Figure 3.3 illustrates the distribution of solutes around a sinking and leaking particle, for example, a marine snow aggregate. Recall that the diffusion–advection equations describe the situation equally well whether the sphere is absorbing or leaking solutes. Depending on the

Péclet number, the solute plume can be very elongated. Solute plumes trailing behind sinking marine snow aggregates may represent volumes of water orders of magnitude larger than the volume of the particles themselves where the concentrations of amino acids, for example, are significantly elevated above background concentration (fig. 3.7). Organic solutes leak from marine aggregates because of the activity of bacteria that have colonized the aggregate. The bacteria use ectoenzymes to solubilize the particle, and they then take up part of the generated solutes. However, solubilization rates typically exceed the rate at which the bacteria take up the solutes (Smith et al. 1992), and the surplus "production" of dissolved organics produces such plumes. Bacteria colonize not only the falling particle but also the solute plume with elevated concentrations of dissolved organics. Many pelagic bacteria possess chemosensory capabilities (fig. 2.11) that allow them to aggregate in regions with elevated concentrations of dissolved organics (on which they feed) (see e.g., Mitchell et al. 1996, Blackburn et al. 1998), and the pelagic bacteria are consequently heterogeneously distributed, even on very fine scales (Seymour et al. 2000). Simulation and other studies have suggested that the bacterial activity on particles and in the solute plume traveling behind particles may account for the majority of the bacterial activity in the water column, even though particles and plumes represent only a very small fraction (a few percent or less) of the water mass (Kiørboe and Jackson 2001, Azam and Long 2001). Thus, there is substantial heterogeneity in the distribution of dissolved substances in the ocean, generated by both localized and ephemeral events, and this heterogeneity is important for microbial processes in the water column. Obviously, routine measurements of bacterial production in small well-mixed water samples (typically a few milliliters) using radioisotopes to estimate cell division rates will miss this kind of heterogeneity.

3.9 THE CHEMICAL TRAIL: MATE FINDING IN COPEPODS

The solute plume left behind a falling particle may also function as a chemical trail. Many copepods and other crustaceans feed on marine aggregates and may localize them by encountering and following the chemical trails that they leave (fig. 3.8). Such "trail tracking" in pelagic invertebrates was first described for a pelagic shrimp (*Acetes sibogae australis*) that follows the scent trail generated by sinking food particles (Hamner and Hamner 1977). This discovery was prompted by an observation made by the authors while diving at 15 m depth, where they found (and sampled) a "remarkable amorphous material" that was covered by

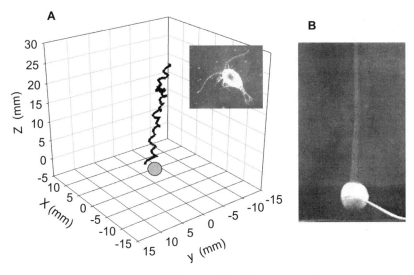

Fig. 3.8. Copepod tracking a sinking particle. The copepod *Temora longicornis* (insert) tracking amino acid trail (A). Invisible amino acids (panel A) or a fluorescent dye (panel B) is slowly pumped into a porous model particle suspended in an upward flow. Copepods that encounter the invisible chemical trail will follow it toward the particle (panel A). The swimming copepod was filmed simultaneously with two cameras positioned at right angles, thus allowing reconstruction of the three-dimensional swimming track shown in panel A. Modified from Kiørboe (2001).

copepods. It turned out that a boat tender had become seasick and vomited from the side of the boat, and the sinking amorphous material had left a scent trail that attracted copepods in huge numbers!

More exciting (but in some sense perhaps less appetizing) chemical trails are left by many female copepods in mating mode. Males encountering such pheromone trails are able to closely follow the trail to the female (Doall et al. 1998, Yen et al. 1998, Tsuda and Miller 1998, Bagøien and Kiørboe 2005). Such trail tracking can be quite spectacular with the male following a convoluted trail for tens of centimeters that had been covered by the female 1/2 min or longer before (fig. 3.9). During tracking, which is very rapid, the male constantly checks the borders of the trail by means of chemoreceptive sensors on the antennules (he cannot see) and adjusts his direction accordingly. This is possible because the borders of the trail are very sharp (fig. 3.3). In contrast, the concentration gradient in the direction of the trail is very weak. Thus, when encountering a trail, how does he know in which direction to find the

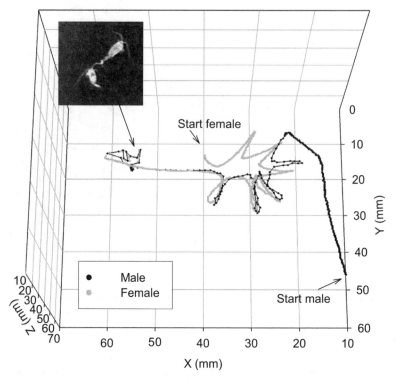

Fig. 3.9. Mate finding in a copepod (*Centropages typicus*). Male copepod tracking pheromone trail of female. Each dot represents position of female (gray) or male (black) at 0.04-s intervals. The female swims in feeding loops while leaving a pheromone trail. The male swims with more directional persistence, which will enhance his chances of encountering a female trail. On encountering the trail he accelerates and faithfully follows the trail through all the loops until he eventually meets the female. He uses his first antennules to seize the female (insert); the couple may remain in precopula for many minutes until he eventually completes his mission and positions a spermatophore on the telson of the female. Modified from Bagøien and Kiørboe (2005).

female? The answer is that he does not. In about half the cases he starts tracking in the wrong direction, but often realizes the mistake after a while, turns around, and makes it to the female (Doall et al. 1998, Bagøien and Kiørboe 2005).

All such cases of generating and tracking chemical trails are high-Péclet-number situations and can be analyzed or described by (variants of) the model presented above (fig. 3.3). However, there is no analytical solution to the model (i.e., we cannot write the solution as an equation). In order to examine the situation in further detail, we simplify the description a bit and treat the target as a moving point source. (We will have a swimming copepod female in mind.) Then the advection part becomes very simple because we simply ignore the effect of the body on the flow, and we can write an analytical solution to the appropriate advection-diffusion equation (modified from Okubo 1980, Jackson and Kiørboe 2004):

$$C_{x,y,z} = \frac{Q}{4\pi Dz} \exp\left(-\frac{ur^2}{4Dz}\right) \tag{3.12}$$

where z is the along-track distance in the swimming or sinking direction, r is the radial distance from the track centerline perpendicular to the track direction, and u the swimming/sinking velocity of the target. Recall also that C is the concentration of the substance (e.g., pheromone) with diffusivity D that is released at rate Q from the point source. (The solution ignores diffusion in the z direction, and we have assumed the background concentration C_∞ to be zero.)

A couple of counterintuitive results immediately fall out of this analysis. First, if we consider solute distribution only along the centerline of the trail (i.e., $r=0$), then equation 3.12 becomes identical to the equation describing the steady-state solute distribution around a nonmoving point source (eq. 2.30, section 2.9). That is, moving does not leave chemical footprints to any further distance than diffusion could reach. This is counterintuitive! The difference is the time required for the signal to reach a certain distance. For the chemical signal to reach a certain distance l from the female, it would take time l/u for the cruising female but l^2/D for the stationary one. These two times would, for typical diffusivities (10^{-5} cm^{2}s^{-1}), copepod swimming velocities (10^{-1} cm s^{-1}), and detection distances (10 cm), be 100 s and 4 months, respectively. Thus, chemical signaling of the type considered here works only for moving targets.

We can estimate the length of the detectable plume, L, by assuming a threshold concentration for detection, C^*. By putting $z=L$, $C=C^*$ and $r=0$ in equation 3.12 we get

$$L = \frac{Q}{4\pi DC^*} \tag{3.13}$$

Note again that, counterintuitively, trail length is independent of female swimming velocity. As a first approximation it is fair to assume that C^* is independent of size in copepods, whereas the leakage rate of pheromones is likely to increase with size. Because trail length is directly

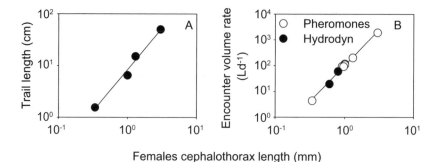

Fig. 3.10. Lengths of detectable copepod pheromone trails (A) and estimates of mate encounter volume rate in pelagic copepods as a function of female size (B). Panel B considers both species that use pheromone and those that use hydrodynamic signals in detection and localization of mates. Data compiled from various sources by Kiørboe (2007).

proportional to leakage rate, we would, thus, expect large copepods to producer longer detectable trails than small copepods, a prediction that is borne out by the few available observations (fig. 3.10A).

Obviously, the purpose of the pheromone trail is to improve the encounter rate between mates. How much the encounter rate is increased by the pheromone trail can be examined by defining the encounter rate kernels for the two situations in which the male encounters the female directly or encounters a pheromone trail. To keep it as simple as possible, consider the female either as a point or as represented by the length of the chemical trail and in both cases moving with velocity u_φ. Similarly, consider the male as an imaginary sphere of radius R (sensory reach, for example, the length of the antennules) moving with velocity u_σ. The encounter-rate kernel for direct encounter in the absence of a pheromone signal is:

$$\beta_{\text{Direct}} = \pi R^2 \Delta u \tag{3.14}$$

where Δu is the velocity difference between male and female, which can be approximated by $(u_\varphi^2 + u_\sigma^2)^{0.5}$. The volumetric encounter rate for the trail tracker is

$$\beta_{\text{Trail}} = 2LRu_{\text{perpendicular}} \tag{3.15}$$

where $u_{\text{perpendicular}}$ is the component of the male swimming velocity that is perpendicular to the direction of the female trail. If the male is swimming in a random direction, then on average $u_{\text{perpendicular}} = 0.82 \times u_\sigma$ (as can be shown from the Pythagorean theorem). Note that the trail encounter rate is independent of female swimming velocity. The enhancement in mate encounter as a result of pheromone production can be estimated by the ratio of the encounter-rate kernels, $\beta_{\text{Trail}}/\beta_{\text{Direct}} \approx 0.5$ L/R. This is because $\Delta u \approx u_\sigma$ when the male swims faster than the female, which is typically the case. If we take body length as an estimate of R, then for the copepod species considered in figure 3.10A, the enhancement in encounter rate increases from a factor of 20 for the smallest species to 85 for the largest; that is, it increases with size. Thus, chemical communication substantially increases mate encounter rates, and most so for large-bodied species.

Absolute magnitudes of estimated volumetric search rates in pelagic copepods are impressive, from several liters per day in the tiniest copepods to several cubic meters per day in the larger—but still small (3 mm)—*Calanus* species (fig. 3.10B). Independent of the kinds of signals used in remote mate detection, the volumetric mate encounter rate increases approximately with the third power of the body length. This makes sense because larger animals are, generally, sparser than smaller ones, and concentrations appear to decline approximately inversely with body mass (and, hence, length^3); such a scaling of volumetric encounter rates is therefore required to allow similar mate encounter rates for small and large copepods.

How expensive is it to produce pheromones? This question is difficult to address because we do not know much about the nature of signal molecules in copepods and other plankters. However, pheromone molecules are likely to be small (Kiørboe and Bagøien 2005), so let us for the sake of this exercise assume that the signal molecules are amino acids.[5] Copepods and other crustaceans are known to respond to amino acids, and sensitivities to concentrations (C^*) down to about 10^{-8} M or less have been recorded in copepods, which is close to the typical background concentration of amino acids in the upper ocean (reviewed in Kiørboe and Thygesen 2001). From equation 3.13, we have that $Q = 4\pi DC^*L$. Assuming typical copepod values of $L = 10$ cm, $C^* = 10^{-8}$M $= 10^{-11}$ mol cm^{-3}, and $D = 10^{-5}$ cm^{2}s^{-1} yields an estimated $Q \approx 10^{-14}$ mol amino acid s$^{-1} = 10^{-14}$ mol N s^{-1}. The typical ingestion rate of a 1.5-mm copepod is on the order of 10^{-12} mol N s^{-1}. Thus, pheromone production apparently represents only a small, but potentially very rewarding, investment.

[5]Oligomers of amino acids that have been demonstrated as chemical cues for settlement of benthic invertebrates (Zimmer-Faust and Tamburi 1994; Brown and Zimmer 2001) would give more specificity to the signal, but diffusive mating signals in copepods appear not to be very specific (Erica Goetze, personal communication), and trail-tracking can be elicited by amino acids in copepods (Kiørboe 2001).

Trail tracking has also recently been described on a much smaller scale, that of marine bacteria tracking trails of swimming phytoplankters (Barbara and Mitchell 2003b). The motile bacteria examined were able to follow a small (5 μm diameter) swimming alga and thus remain within its phycosphere of elevated concentration of organic solutes. Moreover, the bacteria were able to follow trails of algae, much the same way as described above for shrimp and copepods (although the actual mechanism by which the bacteria do this is much different). The time and space scales are, however, much different with detectable trail ages on the order of 0.5 s and trail lengths of about 15 μm. The Péclet number of the swimming algae (65 μm s^{-1}) is only about 0.2, so the distribution of algal exudates around the cell will be determined mainly by diffusion, and thus, the phycosphere is only slightly elongated along the swimming track of the algal cell. Applying the moving point source model (eq. 3.13) and making reasonable assumptions of exudation rate ($Q = 10\%$ of the algal N content per day), diffusivity of the exudates ($D = 10^{-5}$ cm^2 s^{-1}) and respond threshold ($C^* = 10^{-8}$ M) yield trail lengths of the correct order (~50 μm). Chemical signals thus function on a large range of spatial scales in the pelagic environment, but on the small scale of individual phytoplankters, the "need" for advection to spread the signal is much less than at the larger scale of the copepods. As noted previously, this is because diffusion is so fast and efficient at the small scales and so inefficient at the larger scales.

On the small scale of individual phytoplankton and bacteria, viscosity prevents turbulence from interfering with chemical signals. However, at the larger scales of the scent trails of copepods and sinking particles, turbulence may erode the chemical trail. Simulation studies by Visser and Jackson (2004) suggest that with increasing turbulence, the length of the unbroken trail will tend toward half its length in undisturbed water. Some larger copepods appear to aggregate at depth for mating during spring (Tsuda and Miller 1998); this ensures high concentrations of mates (hence high encounter rates) and a quiescent environment where the odor trail remains intact. Small, neritic copepods with multiple generations during summer appear to remain in the surface layer and, thus, have to cope with turbulent erosion—but not destruction—of female trails.

Chapter Four

PARTICLE ENCOUNTER BY ADVECTION

I N THE FOREGOING CHAPTERS we have considered diffusion-type processes and diffusion in combination with advection in bringing particles together. In this chapter we examine processes in which advection dominates and diffusion can be (or at least is) ignored. Advection always implies that water moves relative to the particle or organism considered, either because the object swims or sinks or because of velocity gradients generated in the ambient water (e.g., from turbulence). Two major problems come to mind, namely particle feeding in plankters that generate a feeding current or cruise through the water and formation of marine snow aggregates by physical coagulation. In both cases, particles may encounter one another because they themselves sink or swim and/ or because ambient shear (turbulence) brings them together. When considering these phenomena, we cannot always disregard diffusive process; for example, Brownian diffusion may be important in coagulating very small particles, and even in flagellates that generate strong feeding currents, diffusion can be important (in combination with advection) for encounters with colloidal-sized prey particles. Thus, the topics in these chapters are not completely distinct.

4.1 DIRECT INTERCEPTION VERSUS REMOTE DETECTION

When considering particle encounters, particularly feeding, we distinguish between direct interception of prey particles and remote detection of prey particles. The hydrodynamics is different between the two situations. The presence of a solid body affects the local fluid flow (cf. Stokes' flow around a sphere, fig. 3.1) and, therefore, the rate at which prey particles are intercepted. Many planktonic predators, however, are capable of remotely detecting their prey, and prey are therefore rather encountered by a "perceptive" sphere, which of course does not influence the local flow. In the next chapter we shall examine how remote detection functions by hydrodynamic cues. Here we consider only encounter rates. One can formulate encounter rate kernels for both directly intercepted and remotely detected prey; they are referred to as, respectively, curvilinear and rectilinear kernels. We consider direct interception first.

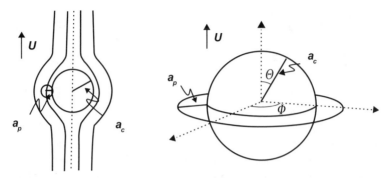

Fig. 4.1. The encounter-rate kernel for direct interception can be computed as the volume flow through the annulus between the collector and the streamline one prey particle radius away from the collector.

4.2 PARTICLE ENCOUNTER BY DIRECT INTERCEPTION: FLAGELLATE FEEDING

Many heterotrophic flagellates generate feeding currents and directly intercept prey particles arriving in the feeding current (Fenchel 1982). This implies that only prey particles that are following streamlines that are within one prey-particle radius (a_p) of the flagellate cell body will be captured (fig. 4.1). The encounter kernel (clearance rate) can therefore be estimated as the volumetric flow rate of water through the annulus between the cell body and the limiting streamline. Can we estimate the magnitude of this flow? As a first approximation, and because flagellates are near-spherical and operate at Reynolds numbers <<1, we can assume Stokes' creeping flow around a sphere (cf. section 3.3, fig. 3.1). The flow velocity, split into its radial and tangential components and expressed using polar coordinates, is given by:

$$u_r = -U \cos\theta \left(1 - \frac{3a_c}{2r} + \frac{a_c^3}{2r^3}\right) \tag{4.1}$$

$$u_\theta = U \sin\theta \left(1 - \frac{3a_c}{4r} - \frac{a_c^3}{4r^3}\right) \tag{4.2}$$

where U is the swimming velocity of the cell, r is the distance to the center of the sphere, θ the angle relative to the direction of the motion of the sphere, and a_c the radius of the cell (the collector). To estimate the volumetric flow rate through the annulus, we need to integrate the flow velocity from the surface of the sphere (where it is zero, according to the no-slip condition) out to one prey radius (a_p), i.e., from a_c to $a_c + a_p$, as

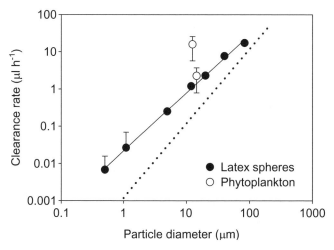

Fig. 4.2. Clearance versus prey size in the dinoflagellate *Noctiluca scintillans*. Filled symbols are clearance rates on spherical latex beads, and open symbols are clearances on phytoplankton cells. The dotted line is the expected scaling (eq. 4.3). *Noctiluca* does not produce a feeding current by means of a flagellum but ascends in the water because it is positively buoyant (Kesseler 1966). It scavenges prey as it ascends, not only on the cell body but also—and probably mainly—on a mucus thread that it produces. Despite its large cell size, ~400 μm diameter, and a consequently low volume-specific clearance rate, the species can survive as a scavenger because it has a very low biomass content per cell volume and, hence, a relatively high mass-specific clearance rate. The volume-specific carbon content of *Noctiluca* cells is about 2 orders of magnitude less that of other dinoflagellates (Nakamura 1998). The inflated size allows for a rapid ascent rate and, hence, high prey encounter rate. Modified from Kiørboe and Titelman (1998).

well as all the way around the equator of the sphere, i.e., from 0 to 2π. Hence:

$$\beta_{\text{Interception}} = \int_{a_c}^{a_c+a_p} \int_0^{2\pi} r u_\theta\left(r, \frac{\pi}{2}\right) dr \cdot d\phi$$

$$\approx \frac{3}{2}\pi U a_p^2$$

(4.3)

(for $a_p \ll a_c$).

One surprising implication of this is that clearance is independent of the size of the predator but increases with prey size squared. This, in turn, implies that the volume-specific clearance rate (i.e., clearance rate divided by cell volume) scales inversely with cell size cubed. Thus, specific clearance decreases dramatically with cell size for interception feeders, and only small predators going for big prey can make a living out of interception feeding. The scaling predicted by equation 4.3 applies to interception feeders generally because it is largely independent of the shape of the collector. Observations of clearance rates in interception feeders conform largely to the predicted scaling with prey size (fig. 4.2). (Recall the request in section 3.6 to compare the efficiency of direct interception with advection–diffusion feeding; you should see that the latter mechanism is much more efficient than the former.)

4.3 BACTERIA COLONIZING PARTICLES REVISITED: COMPARISON OF ENCOUNTER MECHANISMS

The above equation 4.3 has more general applicability in that it describes the rate at which large sinking particles collide with small ones. In chapters 2 and 3 we examined bacterial colonization of particles, such as marine snow aggregates, as examples of diffusive and advective-diffusive transport, respectively, to a spherical collector. A sinking marine snow aggregate also intercepts (scavenges) bacteria as it falls through the water. Because the sinking velocity of a typical marine snow aggregate is much higher than typical swimming velocities of bacteria, intuition would suggest that particle sinking and scavenging are much more important than bacterial motility in bringing particles and bacteria together. This is, however, far from being the case. Comparing the encounter kernels for diffusion (eq. 2.22), advection (eq. 4.3), and advection+diffusion (eq. 3.4) reveals that diffusion (bacterial motility) is orders of magnitude more important than scavenging in causing bacterial-particle encounters (fig. 4.3). Advection is significant for large particles, but only together with bacterial motility. This implies that nonmotile bacteria hardly colonize even sinking particles. This finding is consistent with the finding above, that direct interception is a very inefficient way for large particles to get in contact with small ones. The reason is the hydrodynamics: the larger, sinking particle pushes water (and bacteria) away as it falls.

4.4 DIRECT INTERCEPTION: COAGULATION AND MARINE SNOW FORMATION

Equation 4.3 may also be used to describe the rate at which larger and faster-sinking particles encounter smaller and slower particles during

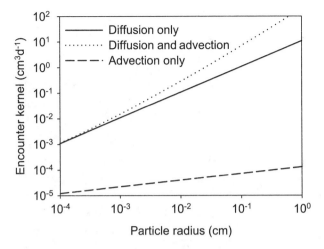

Fig 4.3. Comparison of encounter kernels as a function of particle size for bacterial colonization of a particle by bacterial diffusion alone, by advection (direct interception) alone, and by the combined effect of advection and diffusion. The three scenarios correspond to motile bacteria colonizing a stationary particle, nonmotile bacteria colonizing a sinking particle, and motile bacteria colonizing a sinking particle, respectively. The computations have been made from equations 2.22, 3.4, and 4.3, assuming a bacterial diffusivity of 10^{-5} cm^2 s^{-1}, bacterial diameter of 0.5 μm, and a particle sinking velocity (u, cm s^{-1}) as a function of particle radius (a, cm) given by $u = 0.13a^{0.26}$ characteristic of marine snow aggregates (Alldredge and Gotschalk 1988).

descent,[1] but the sinking velocity U should then be replaced by the velocity difference, Δu. Particles that collide with one another may stick together, either by electrostatic forces, because they have surfaces covered by mucus that glues particles together, or because they become mechanically entangled (e.g., diatoms with long chitin threads). Subsequent and continued collisions and sticking of small particles may lead to the formation of larger aggregates, known as marine snow (fig. 4.4). Such aggregates can consist of all kinds of small particles—phytoplankton cells, fecal pellets, zooplankton exuvia, etc.—and they may vary in composition (fig. 4.4). Aggregates can be very important both because they may account for substantial vertical material transport in the ocean and because

[1]Equation 4.3 is, except for the lead coefficient, identical to the coagulation kernel for differential settling in the classical coagulation literature (with U replaced by velocity difference, Δu) (Pruppacher and Klett 1978). The classical derivation has a different lead coefficient, 0.5 instead of 1.5. The reason is that the classical derivation assumes that both the large and the small particle sink according to Stokes' law and that they have identical densities.

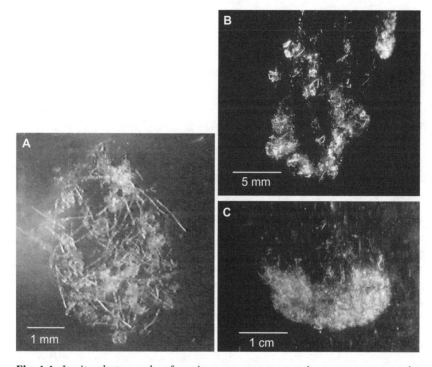

Fig. 4.4. In situ photographs of marine snow aggregates. Aggregates can consist of all kinds of small particles, living as well as dead. The aggregate in A consists almost entirely of chains of the diatom *Chaetoceros* sp. Aggregate B consists of a mixture of particles, some of which can be identified as diatom cells, others as zooplankton fecal pellets. Finally, aggregate C consists again mainly of diatoms. B and C by courtesy of Alice Alldredge. From Kiørboe (2003), reproduced with permission of Inter Research.

they are microbial hot spots. We shall therefore consider marine snow formation by physical coagulation in a bit more detail.

For simplicity, let us consider a bloom of uniformly sized phytoplankters of radius a_1 and concentration C_1. The encounter rate or collision frequency between suspended cells is then (from eq. 1.1a)

$$E = \beta C_1 C_1 = \beta C_1^2 \tag{4.4}$$

If the cells stick together on collision with a certain probability, α, also called particle stickiness, then the concentration of single (unaggregated) cells will change according to

$$\frac{dC_1}{dt} = -\alpha \beta C_1^2 \tag{4.5}$$

TABLE 4.1
Curvilinear Encounter Kernels

Mechanism	Encounter Rate Kernel (L^3T^{-1}) $\beta_{i,j}$	Equation Number
Brownian diffusion	$4\pi(D_i+D_j)(a_i+a_j)$	4.6
Differential settling	$0.5\pi a_i^2\,\lvert u_i-u_j\rvert$ for $a_i \le a_j$	4.7
Small-scale turbulent shear	$1.3\gamma(a_i+a_j)^3 E_{i,j}$ where $E_{ij}=1$ for $a_i=a_j$ $$E_{ij} = \frac{7.5(a_i/a_j)^2}{[1+2(a_i/a_j)]^2}\ \text{for } a_i < a_j$$	4.8

Curvilinear encounter-rate kernels (coagulation kernels) between spherical Particles of radii a_i and a_j. The small-scale turbulent shear rate is estimated as $(\varepsilon/v)^{0.5}$, where ε is the turbulent energy dissipation rate and v is the kinematic viscosity. By "small-scale" is meant scales smaller than the Kolmororov length scale $=(v^3/\varepsilon)^{0.25}$ (see section 4.6). Coefficients for Brownian diffusion can be computed from eq. 2.26. Based on Jackson (1990) and Hill (1992).

As usual, β is the encounter-rate kernel. Particles may encounter one another by various mechanisms, namely by thermally driven Brownian diffusion, by faster sinking particles overtaking slower sinking ones (differential settling, cf. above), and by ambient fluid motion (turbulence) bringing particles into contact (fig. 4.5). For each of these processes, we can write an encounter-rate kernel (table 4.1). We are already familiar with the kernels for diffusion and differential settling; the slight differences between the equations given in table 4.1 and previously are that here we need to consider the sizes, velocities, and diffusivities of both of the particles colliding.

In the engineering literature it is frequently assumed that the β that goes into equation 4.5 is the sum of βs of the relevant processes. We shall follow this practice here, but we note that the alternative Sherwood-number approach that we have used previously is more accurate. For the problem at hand, we can disregard Brownian motion because it becomes unimportant for particles larger than about 1 μm, i.e., for most phytoplankters (as can be seen from eq. 2.26). Also, because we are considering a monospecific phytoplankton bloom, all cells sink with about the same velocity; hence, the kernel for differential settling vanishes. We are thus left with turbulent shear as the only significant process, at least initially. Thus,

$$\frac{dC_1}{dt} = -\alpha\beta C_1^2 = -\alpha 1.3\gamma\left(a_1 + a_1\right)^3 C_1^2 = -\alpha 10.3\gamma a_1^3 C_1^2 \tag{4.9}$$

As coagulation proceeds, aggregates consisting of two, three, four, . . . cells begin to form, and collisions between single cells and small

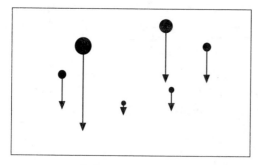

Differential settling

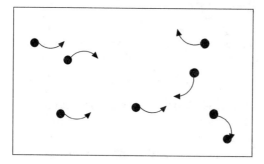

Turbulence

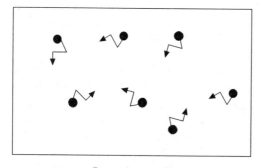

Brownian motion

Fig. 4.5. Schematic of physical particle collision mechanisms: differential settling, turbulence, and Brownian motion.

aggregates and between aggregates of various sizes will occur. One can keep track of all possible collisions and aggregate sizes by infinitely many coupled differential equations (see Jackson 1990 and Jackson and Lochman 1992 for ways to handle this), but it can be shown that, if one considers only the initial process, a good approximation is (Kiørboe et al. 1990)

$$C_t = C_0 \exp[-\alpha(7.8\phi\gamma/\pi)]t \qquad (4.10)$$

where C_0 is the initial concentration of single cells, C_t is the total concentration of particles (single cells + aggregates) at time t, and ϕ is the volume fraction of cells $(=4/3\pi a_1^3 C_0)$. Likewise, the average solid volume of particle aggregates will increase according to

$$V_t = V_0 \exp[\alpha(7.8\phi\gamma/\pi)]t \qquad (4.11)$$

where V_t and V_0 are the average particle volumes at times t and 0 (V_0 of course is the volume of a single phytoplankton cell). Thus, as particles combine through coagulation, particle concentration will decline, and average particle size will increase, both of them exponentially. An experimental demonstration that phytoplankton cells in fact can form aggregates by coagulation is given in figures 4.6 (qualitative) and 4.7 (quantitative).

The significance of aggregation in the ocean is severalfold. Aggregation has important implications for the dynamics of phytoplankton populations (chapter 7) and for the structure and function of pelagic food webs (chapter 8). Aggregates occur in all oceans at all times and often at high concentrations. Because particle sinking velocity is proportional to particle radius squared and to the density difference between the particle and the ambient fluid (Stokes' law), aggregates sink faster or much faster than the component particles. Small algal cells, for example, sink at rates of some centimeters per day, whereas large aggregates may sink at rates of up to hundreds of meters per day. Aggregates therefore account for a significant fraction of the vertical transport in the ocean (Fowler and Knauer 1986), and vertical sinking fluxes can often be accurately predicted from the simple theory presented here (e.g., Kiørboe et al. 1994, Jackson et al. 2005; fig. 4.8). Aggregates are also concentrated packages of particulate organic matter. They therefore represent feeding opportunities for a variety of organisms that would otherwise have to concentrate organic particles from the meager food suspension that seawater is. Copepods, other metazoans, protozoans, and bacteria all colonize aggregates and feed on the component particulate material, and abundances of heterotrophs in or on aggregates are typically many orders of magnitude higher than in the ambient water. Thus, aggregates are heterotrophic hot spots where significant remineralization of organic material occurs. For more information on marine snow aggregates, consult reviews by Alldredge and Silver (1988) and Simon et al. (2002) and check publications by G. A. Jackson.

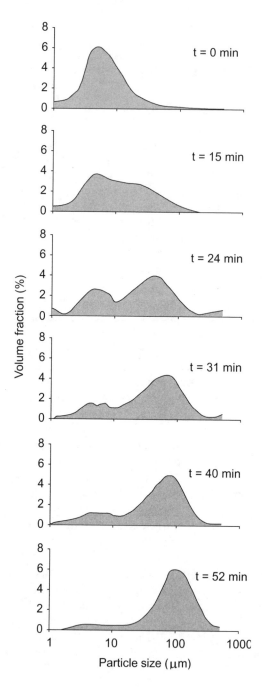

Fig. 4.6. Qualitative demonstration of phytoplankton aggregation by coagulation. Phytoplankton cells may aggregate when suspended in a sheared flow field. A suspension of diatoms, *Skeletonema costatum* (equivalent spherical diameter ~5 μm), was exposed to strong laminar shear ($30 \ s^{-1}$), and the temporal development of the particle size distribution was monitored over time using laser diffraction. Aggregates were gradually formed during the incubation (Kiørboe 1997).

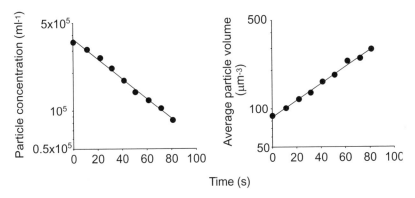

Fig. 4.7. Quantitative demonstration of phytoplankton aggregation by coagulation. Phytoplankton cells may aggregate in a turbulent environment as predicted by classical coagulation theory. Diatoms, *Phaeodactylum tricornutum* (~5 μm) were suspended ($\varphi = 33 \times 10^{-6}$) in a beaker with an oscillating grid generating turbulence with a dissipation rate of $\varepsilon = 25\,cm^2 s^{-1}$ (equivalent to a shear-rate, γ, of 50 s^{-1}). Total particle concentration declines, and average particle volume increases, exponentially as predicted by equations 4.10 and 4.11. Because the only unknown in equations 4.10 and 4.11 is the stickiness, α, it can be estimated from the slopes (in this case $\alpha = 0.12$ is estimated). Modified from Kiørboe et al. (1990).

4.5 REMOTE PREY DETECTION: ENCOUNTERING PREY IN CALM WATER

We return now to predators encountering prey but this time consider situations where the predator can remotely detect the prey. Most fish larvae use vision to remotely locate prey, but most copepods use either chemical or hydrodynamic cues to remotely detect prey (see next chapter). Even copepods that generate a feeding current and apparently filter prey particles out of the water may remotely detect the arriving prey and only filter the small parcel of water actually containing that prey (Koehl and Strickler 1981; see nice account in chapter 2 of Mann and Lazier [1991]). If the encounters between prey and predator are not by direct interception but rather so that the prey comes within the perceptive (or dining) sphere of the predator, then, of course, there is no impact on the fluid motion by this imaginary sphere. For a predator that moves (cruising, sinking) or generates a feeding current, the encounter rate kernel is simply (fig. 4.9):

$$\beta_{Cruising} = \pi R^2 u \tag{4.12}$$

where u is the swimming (or sinking) velocity or feeding current velocity of the predator, and R the distance at which the predator perceives

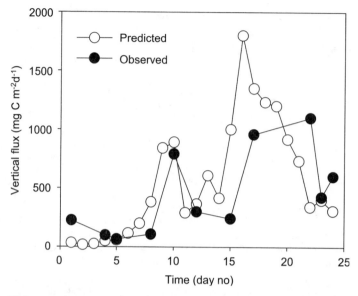

Fig. 4.8. Observed and predicted vertical flux of phytoplankton during a diatom spring bloom in a Danish fjord. The vertical flux was predicted from simple co-agulation theory as $\alpha\beta C^2$ (cf. equation 4.5), considering only collisions between unaggregated cells. Even when aggregates are formed, this is the dominant process. The parameters used for the prediction were estimated from daily measurements of phytoplankton concentrations (C), coagulation experiments (α), and from wind-based estimates of turbulent shear (β). Modified from Hansen et al. (1995).

and responds to the prey (to this should be added the radius of the prey).

If the prey also moves along a more or less straight path, then u should be replaced by $(v^2 + w^2)^{0.5}$, where v and w are predator and prey swimming velocities. An example of the application of equation 4.14 is the copepod *Acartia tonsa* feeding on ciliates while in ambush feeding mode (sinks steadily at $v = 0.67\,\text{mm s}^{-1}$). *A. tonsa* can perceive swimming ciliates at a distance of about 0.1 cm by the hydrodynamic signal that they generate; thus, $R = 0.1$ cm. Equation 4.14 thus predicts a clearance rate of 187 ml d^{-1}, which is very close to the clearance rate actually measured (182 ml d^{-1}; Saiz and Kiørboe 1995). (The swimming velocity of the ciliate, ~0.1 mm s^{-1}, makes an insignificant contribution to the clearance rate.) Of course, there are other motility patterns than sinking or swimming along more or less straight paths, and encounter

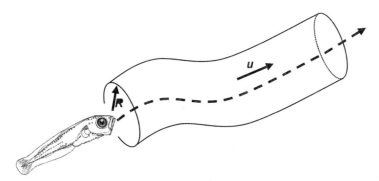

Fig. 4.9. The encounter-rate kernel for a predator cruising at velocity u and perceiving all prey within a distance R can simply be computed as the volume of the "perceptive tube." This is the volume of water that the predator searches for prey per unit time.

kernels can be derived for these situations. For example, many fish larvae can be considered "pause-travel" predators: the predator moves a short distance, stops, searches for prey within the dining sphere, and then moves to the next stop. This search strategy is also known for many birds, e.g., thrushes that search the lawn for worms. The typical move distance is similar to or slightly larger than $2R$; that way non-overlapping volumes of water are searched, and move distances are kept to a minimum. The kernel is given in table 4.2, and an example of its application to larval cod may be found in MacKenzie and Kiørboe (1995).

4.6 TURBULENCE AND PREDATOR-PREY ENCOUNTER RATES

The predator–prey encounter-rate kernels dealt with until now can all be considered behavioral kernels. But encounters between predator and prey may also occur as a result of ambient fluid motion. The idea that turbulence may enhance the contact rate between planktonic predators and their prey was first introduced by Rothschild and Osborn (1988). Thus, we can also define kernels caused by ambient turbulence, *turbulence kernels*. Kernels related to behavior and those from fluid motion are additive (as a good approximation). Comparing magnitudes of kernels from behavior and those from turbulence offers a good way to evaluate theoretically the potential effect of turbulence on encounter rates. We first spend some time discussing what

TABLE 4.2
Predator-Prey Encounter Kernels

Predator Behavior and Mechanism	Encounter Rate Kernel $(L^3 T^{-1})$ $\beta_{i,j}$	Equation Number
Cruising, sinking, feeding current	$\pi R^2 u$	4.12
Pause-travel	$\frac{4}{3} \pi R^3 f$	4.13
Random walk (diffusion)	$4\pi D R = \frac{4}{3} \pi u^2 \tau R$	4.14
Turbulence <Kolmogorov scale	$1.3 \gamma R^3$	4.15
Turbulence >Kolmogorov scale	$1.37 \pi R^2 (\varepsilon R)^{1/3}$	4.16

Behavioral and physical rectilinear kernels for predator-prey encounter rates. The velocity u in eq. 4.12 is the velocity difference between predator and prey, which can be approximated by $(v^2 + w^2)^{0.5}$, where v and w are predator and prey swimming or feeding current velocities. f in eq. 4.13 is the stop frequency. The two versions of eq. 4.14 utilize the fact that diffusivity can be estimated as $D = u^2 \tau / 3$. Note that γ is the sub-Kolmororov shear rate, which can be estimated from the turbulent energy dissipation rate (ε) as $\gamma = (\varepsilon / v)^{0.5}$, where v is the kinematic viscosity of the water. All equations refer to spherical perceptive fields and may be modified to describe other geometries.

turbulence is and presenting kernels for turbulence, then present a few examples of turbulence-enhanced prey encounter rates, and finally consider some negative effects of turbulence (postencounter processes).

What is turbulence? Turbulence can be generated by wind, tides, currents, and any other processes that add kinetic energy to the water column. Turbulence generated from large-scale phenomena first appears as large-scale eddies. These subsequently dissipate into smaller and smaller eddies, and the smallest eddies eventually dissipate as heat because of viscosity (fig. 4.10).

Although water motion is random, the eddies have a characteristic size distribution, and there is a minimum size of eddies, related to the Kolmogorov length scale, below which fluid viscosity dominates and all turbulent energy dissipates as heat. The rate of energy dissipation, ε ($1 \, W \, kg^{-1} = 10^{-4} \, cm^2 s^{-3}$), is of course equal to the rate at which kinetic energy is added from wind, tides, and other processes. It can be measured in the ocean and is used as a measure of the intensity of the turbulence. Typical dissipation rates in the ocean range from $10^{-1} \, cm^2 s^{-3}$ in the upper ocean during windy conditions to $10^{-5} \, cm^2 s^{-3}$ or less in the ocean interior.

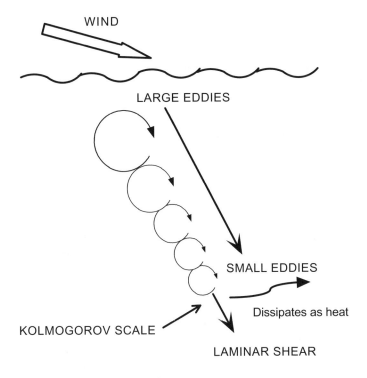

Fig. 4.10. Schematic of turbulence.

In accord with intuition, the size of the smallest eddies ($\sim\eta$ =Kolmogorov length scale) decreases as the energy dissipation rate (=input rate of energy) increases and increases with the viscosity of the fluid; from dimensional analysis,[2] then

$$\eta \approx \left(\frac{v^3}{\varepsilon}\right)^{0.25}$$ (4.17)

[2] By combining (multiply, divide, raise to some power) the parameters that we know impact the property of interest, in this case the Kolmogorov length scale, in such a way that the dimensions become the same on both sides of the equation, we can rationalize the form of the dependency. This is dimensional analysis. The magnitude of the lead coefficient (not indicated in eq. 4.17) is unknown from this type of analysis but is often, sometimes incorrectly, assumed to be of order 1.

For a "typical" upper-ocean dissipation rate of 10^{-2} cm^{2}s^{-3}, $\eta \approx 0.1$–0.5 cm if one assumes a lead coefficient of somewhere between 1 and 2π. Mesozooplankters are, thus, of sizes similar to η.[3]

Turbulence implies that there is a velocity difference between two points in the fluid. If the planktonic organisms are embedded in the fluid flow, then there is also a velocity difference between predator and prey that may lead to the two encountering one another. At scales smaller than the Kolmogorov length scale, turbulence becomes manifested as laminar shear, and the velocity difference, u, simply increases with the separation distance, d, between the points (e.g., Hill et al. 1992):

$$u = 0.42\gamma d = 0.42\left(\frac{\varepsilon}{v}\right)^{0.5} d \text{ for } d \ll \eta \tag{4.18}$$

At scales larger than the Kolmogorov length scale, the velocity difference scales with energy dissipation and separation distance both raised to a power of 1/3, thus (e.g., Hill et al. 1992),

$$u = 1.37(\varepsilon d)^{1/3} \quad \text{for } d \gg \eta \tag{4.19}$$

Now we can formulate the turbulence kernels by using the usual approach of $\beta_{turbulence} = \pi d^2 u$; by setting d = reaction distance (R), and with u given by equation 4.18 or 4.19, we get the encounter-rate kernels listed in table 4.2 (eqs. 4.15 and 4.16).

The equations are valid for spatial scales much larger or much smaller than the Kolmogorov length scale, but the relations are not well known for scales close to η. This situation is unfortunate because most mesozooplankters are of about that size and because turbulence is potentially most important for organisms of that size (see below). However, experiments have suggested that the $d \gg \eta$ equations are good approximations down to and even below η (Hill et al. 1992).

4.7 Example: Feeding of the Copepod *Acartia tonsa* in Turbulence

The effects of turbulence on copepod feeding were examined theoretically by Kiørboe and Saiz (1995) and experimentally by several authors,

---

[3] As an aside: It has recently been suggested (Huntley and Zhou 2004, Kunze et al. 2006) that swimming zooplankters on a local scale may add substantial kinetic energy to the water and that this may help stir the ocean and, e.g., mix nutrients across the thermocline. However, because turbulence cascades from larger to smaller scales (Fig. 4.10), kinetic energy generated by zooplankters at a spatial scale near the Kolmogorov length scale will almost immediately dissipate as heat rather than contribute to the mixing of the ocean (Visser 2007).

AMBUSH FEEDING   SUSPENSION FEEDING

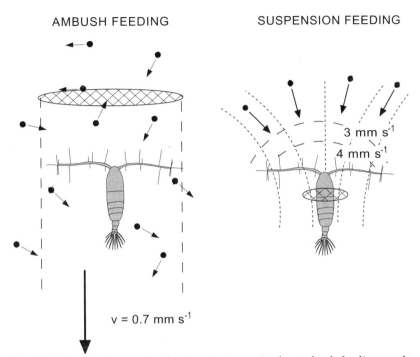

$v = 0.7$ mm s^{-1}

Fig. 4.11. Two feeding modes for *Acartia tonsa*. In the ambush feeding mode, the copepod hangs motionless in the water with the antennae extended while sinking slowly. Motile prey are perceived by hydromechanical receptors on the antennae. In the suspension-feeding mode, the copepod generates a feeding current, and prey that come within the hatched "capture volume" are captured. Sinking velocity of the ambush-feeding copepod (left panel), and the velocity isolines of the accelerating feeding current have been indicated on the drawings (velocity isolines in right copepod; the dashed lines with contour values shown). Data taken from Jonsson and Tiselius (1990).

all of whom demonstrated elevated prey encounter or feeding rates at moderate levels of turbulence (Costello et al. 1990, Marrasé et al. 1990, Saiz and Kiørboe 1995, Caparroy et al. 1998, Saiz et al. 2003). As an example, consider the copepod *Acartia tonsa* (Saiz and Kiørboe 1995). This copepod has two feeding modes: a suspension feeding mode in which it generates a feeding current and captures particles that pass within the capture volume and an ambush feeding mode in which it sinks slowly and perceives and attacks prey passing within the reaction distance (fig. 4.11). The two feeding modes can be triggered by offering the copepod either motile prey (ambush mode) or diatoms (suspension feeding mode) (Jonsson and Tiselius 1990). Above we considered the calm-water prey encounter rate for the copepod in ambush feeding mode using equation 4.12, and we found a good correspondence between observed and predicted feeding rates. If the copepod is feeding in turbulence, its clearance rate is enhanced as predicted in a general way by the theory (table 4.3). To predict the clearance rate in turbulent water, one has to add the behavioral and physical kernels. Above we estimated the behavioral kernel to be 187 ml d^{-1}. At a dissipation rate of 4×10^{-3} cm^{2}s^{-3}, for example, the Kolmogorov scale is about 0.1 cm, i.e., similar to the sensory reach of the copepod. We thus use equation 4.16 to estimate the turbulence-generated encounter kernel: $\beta_{\text{turbulence}} = 1.37\pi R^2(\varepsilon R)^{1/3} = 284$ ml d^{-1}. Because the kernels are additive, $\beta = \beta_{\text{turbulence}} + \beta_{\text{behavior}} = 464$ ml d^{-1}. This again is close to that observed (table 4.3). Note that the observed and predicted effects of turbulence are substantial. However, whereas the predicted effect of turbulence keeps increasing with increasing turbulence, the observed effect declines again at higher (and unrealistic) intensities (see also below).

In the suspension-feeding mode *A. tonsa* captures particles that come inside the capture volume (fig. 4.11), which reaches only about 0.02 cm (R). Feeding-current velocity is about 0.8 cm s^{-1} (Tiselius and Jonsson 1990). Thus, we predict a stillwater clearance rate on diatoms of about 90 ml d^{-1}. This is close to the clearance rate actually measured (60–80 ml d^{-1}). Because of the small R and the large u, the predicted effect of turbulence is small, e.g., a total clearance of 104 ml d^{-1} at 2.3×10^{-2} cm^2 s^{-3} or an increase of ~10%. Observations confirm that there is no measurable positive effect of turbulence in suspension-feeding *A. tonsa*. The velocity difference between predator and prey in suspension–feeding *A. tonsa* is so high that the additional contribution of turbulent fluid motion is small. Thus, behavior is important in assessment of the role of turbulence.

4.8 When Is Turbulence Important for Enhancing Predator–Prey Contact Rates?

As is evident from the example above, the significance of turbulence must differ among species. From intuition we would expect that very large predators, fish for example, would benefit little from turbulence because their swimming speed far exceeds turbulent velocities. Likewise, very small organisms should benefit little because at small scales viscosity dampens turbulent fluid motion.

We can be more explicit by asking at what dissipation rate the kernel for turbulence equals the kernel for behavior; we can term this the critical dissipation rate. Turbulence is more important than behavior for prey

TABLE 4.3
Observed and Predicted Clearance Rates of *Acartia tonsa* in Calm and
Turbulent Water

Dissipation rate cm^2s^{-3}	Observed clearance cm^3d^{-1}	Predicted clearance cm^3d^{-1}
0	182±15	187
4.0×10^{-3}	565±78	464
2.3×10^{-2}	715±141	695
8.6×10^{-1}	315±117	1892
6.9×10^0	318±61	3595
3.7×10^1	269±71	6142

From Saiz and Kiørboe (1995).

encounters only at dissipation rates exceeding the critical dissipation rate. It can be found by solving $\beta_{turbulence}\ (\varepsilon_{Cr})=\beta_{behavior}$ for ε_{Cr}. As an example, let us consider cruising predators. At scales exceeding the Kolmogorov length scale, equate equations 4.12 and 4.16 in table 4.3, and at scales smaller than the Kolmogorov scale, equate equations 4.12 and 4.15. Solving for ε_{Cr} yields

$$R > \eta : \varepsilon_{Cr} = 0.4\frac{u^3}{R} \qquad (4.20)$$

$$R < \eta : \varepsilon_{Cr} = 6\upsilon\frac{u^2}{R^2} \qquad (4.21)$$

The form of these equations suggests that the critical dissipation rate is at a minimum at scales near the Kolmogorov length scale (fig. 4.12). This means that for organisms of that size, only moderate turbulence intensities are required for turbulence to be important for prey encounters. For organisms much smaller or much larger than the Kolmogorov length scale, only unrealistically high dissipation rates enhance predator–prey contact rates. Thus, the general conclusion is that it is only organisms in the size range close to the Kolmogorov length scale that potentially benefit from turbulence.

4.9 ON THE DOWNHILL SIDE: NEGATIVE EFFECTS OF TURBULENCE ON PREDATOR-PREY INTERACTIONS

Although theory and observations suggest that turbulence may enhance predator-prey contact rates, turbulence may have negative effects as well. In the example above with the copepod ambush feeding in

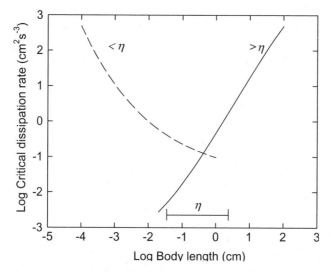

Fig. 4.12. Critical dissipation rate as a function of body size calculated for cruising predators using equations 4.21 and 4.22 and employing the body size–swimming velocity relationship of Peters et al. (1994). R is taken as $1/2 \times$ body length. η is the Kolmogorov length scale, and the typical variation in the magnitude of the Kolmogorov length scale in the upper ocean is indicated. Dissipation rates exceeding 10^0 cm^2 s^{-3} in the upper ocean are rare. Modified from Kiørboe (1997).

turbulence, it was evident from observations that moderate intensities of turbulence enhanced feeding rates, but with a further increase in turbulence, feeding rate began to decline (table 4.3). Thus, there appears to be a dome-shaped relationship between feeding rate and turbulence, as has also been documented in other planktivorous predators, including other copepods (Caparroy et al. 1998, Saiz et al. 2003) and larval fish (MacKenzie and Kiørboe 2000). I shall briefly mention three factors that may cause the decline in feeding rate with increased turbulence:

First, encountered prey may be advected out of the perceptive sphere of the predator before it reacts. The risk that this happens increases with both the dissipation rate and with the reaction distance (both yielding higher relative velocities) as well as with the reaction time. Several models have been developed to describe this phenomenon (e.g., Granata and Dickey 1991, MacKenzie et al. 1994, Jenkinson 1995, Kiørboe and Saiz 1995, Lewis and Pedley 2001). For *Acartia* with a short reaction time (< 0.1 s), this does not become a problem until the dissipation rate exceeds by orders of magnitude what can be found in nature. For larval cod with long prey reaction times (2–4 s), on the other hand, the problem is real at realistic dissipation rates, and it can be demonstrated experimentally (MacKenzie and Kiørboe 2000).

Second, turbulence may erode copepod feeding currents (Jiménez 1997). One can evaluate the magnitude of this potential problem by comparing the magnitude of the velocity difference between the copepod and the ambient water that results from turbulence (eq. 4.19) and the velocity difference caused by the feeding current (fig. 4.11). The latter, of course, decreases with the distance to the copepod, whereas the former increases. At some distance, they are similar. We shall refrain from a formal analysis here but just highlight the conclusion that feeding currents may be significantly eroded at very high dissipation rates. This may be the reason that many copepods appear to migrate downward during periods of strong, wind-generated turbulence in the surface layer (Mackas et al. 1993, Incze et al. 2001).

Finally, turbulence may interfere with hydromechanical prey perception. Copepods and many other planktonic predators perceive swimming prey by means of the hydrodynamic disturbance that the prey generates. This signal may be destroyed by turbulence. We shall consider hydrodynamic signals in the next chapter. Briefly, however, the fluid velocity (signal) generated by a swimming ciliate attenuates with distance as r^{-2}. We assume that a self-propelled organism can be modeled as a force dipole (see chapter 5) and that the velocity difference between the copepod and the ambient fluid varies as $\varepsilon^{1/3}$ (eq. 4.19). Assume that the signal ($\sim r^{-2}$)-to-noise ($\varepsilon^{1/3}$) ratio has to exceed some threshold value K; then (Saiz and Kiørboe 1995, Visser 2001)

$$\frac{r^2}{\varepsilon^{1/3}} = K \Rightarrow r = K\varepsilon^{-1/6} \tag{4.22}$$

Thus, reaction distance is expected to decline with turbulent dissipation rate to the power of $-1/6$. Observations of reaction distance versus dissipation rate in *Acartia* are in surprisingly good agreement with the predicted scaling (fig. 4.13).

4.10 ENCOUNTER RATES AND MOTILITY PATTERNS: BALLISTIC VERSUS DIFFUSIVE MOTILITY

We noted in the introduction to this chapter that the distinction we have made through chapters 2–4 between encounters caused by diffusion, advection, or diffusion and advection is somewhat arbitrary. The insufficiency of this distinction becomes particularly evident in considering organism motility patterns: what may appear as random-walk motility

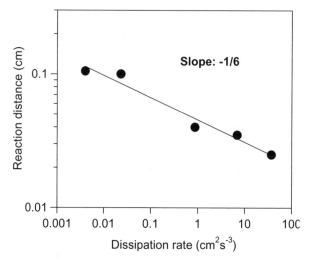

Fig. 4.13. Reaction distance versus turbulent dissipation rate in *Acartia tonsa* preying on ciliates. Data computed from Saiz and Kiørboe (1995).

("diffusive") at one spatial scale may look like directionally persistent cruising at another (advective or "ballistic" motility). Consider as an example a bacterium with a run–tumble motility pattern. At temporal scales less the run duration, τ (or spatial scales less than the run length, $u \times \tau$), such a bacterium appears as a cruiser; at larger spatial and temporal scales, it appears as a random walker. You may recall the two strains of bacteria that we considered in figures 2.1 and 2.2: both had run-tumble motility patterns, but their run durations were very different, 0.4 versus 15 s. When examining their swimming patterns under a microscope preparation, we were able to follow individual bacteria for only 3–4 s. The analyses of the root-mean-square net displacement for the population of bacteria as a function of time showed that the frequent tumbler nicely followed the expected square-root function, whereas the rare tumbler did not: its root-mean-square net displacement increased with time raised to a power much higher than 0.5. In fact, as one further decreases the scale of observation, the power will approach 1, equivalent to cruising along a more or less straight path. Similarly, had we been able to extend the observation time of individual bacteria to much longer than the 15-s average run duration, we would eventually have seen the square-root law fulfilled. Thus, our characterization of an organism's motility pattern is strongly scale dependent: at small scale it appears ballistic, and at larger scales it is diffusive. This idea of a scale-dependent characteristic of the motility pattern is illustrated in fig. 4.14.

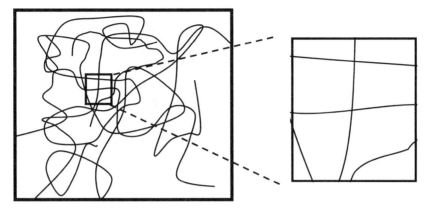

Fig. 4.14. Scale-dependent characterization of motility patterns. Schematic of two-dimensional projections of swimming tracks observed at low magnification (left) and higher magnification (right). At the low magnification, swimming tracks look random, whereas at high magnification the organisms appear to swim along almost straight paths.

The mathematician G. I. Taylor addressed this problem of scale-dependent transition from ballistic to diffusive motility in the 1920s and derived an equation that describes how the root-mean-square distance (*RMS*) increases with time (see Visser and Kiørboe 2006 for details)

$$RMS(t) = \{2u^2\tau[t - \tau(1 - e^{t/\tau})]\}^{0.5} \tag{4.23}$$

where τ is the decorrelation time scale or, in our terminology, the (equivalent) run duration. At small time scales ($t \ll \tau$) *RMS* increases linearly with time; at longer time scales ($t \gg \tau$), it increases with the square root of time. That such a scale-dependent transition from ballistic to diffusive motility in fact applies to organism swimming patterns and that it can be described by Taylor's equation has been exemplified by the motility of a small, heterotrophic flagellate (fig. 4.15).

This scale dependence of the motility pattern has important implications for encounter rates. These can be seen by comparing the encounter kernels for diffusive versus ballistic (cruising) motility (table 4.2); they yield different encounter rates. But if an organism's motility can be characterized both as diffusive and ballistic, then which encounter kernel is the correct one? That depends on the (sensory) size of the target (*R*) relative to the run length (motility length scale = $u\tau$): if the spatial extension of the target is large relative to the motility length scale, then the motility "looks" diffusive (from the point of view of the target), and the diffusion encounter kernel applies (eq. 4.14). Conversely, if the motility length

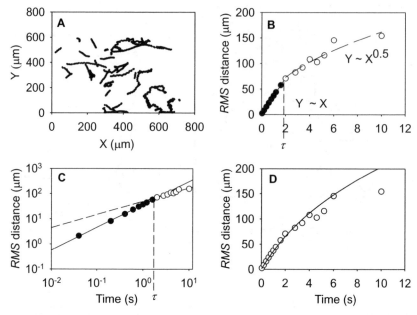

Fig. 4.15. Motility pattern of the 5-μm heterotrophic flagellate *Bodo designis* recorded in a microscope preparation. Two-dimensional projection of selected swimming tracks (A) and root-mean-square (*RMS*) distance traveled as a function of time plotted on linear scales (B, D), and log-log scales (C). In B, a linear and a square-root function, respectively, have been fitted to the data for times less than and larger than 2 s; in the log-log plot (C) the slopes of the two lines are 1 and 0.5, respectively. The average run duration (decorrelation time scale) is at the intersection of the two lines. In D, Taylor's equation has been fitted to the data, and an average run duration of 1.6 s estimated.

scale is small relative to the spatial extension of the target, then the encounter kernel for cruising behavior applies (eq. 4.12).

Ballistic motility is more efficient for encounters than diffusive motility. This is obvious from intuition: a very convoluted swimming track may lead to the searcher scanning the same volumes of water repeatedly, whereas swimming along a straight path takes the searcher constantly though new water. The latter is obviously more efficient. It also follows quantitatively by comparing the kernels for cruising and diffusive motility (eqs. 4.12 and 4.14): when $R > 4/3u\tau$, the encounter kernel for cruising exceeds that for diffusion. This realization leads to the prediction of an optimum motility pattern for an organism that feeds on smaller prey and is itself the prey of larger predators: it should have a run length longer than the distance at which it detects its prey and shorter than the sensing

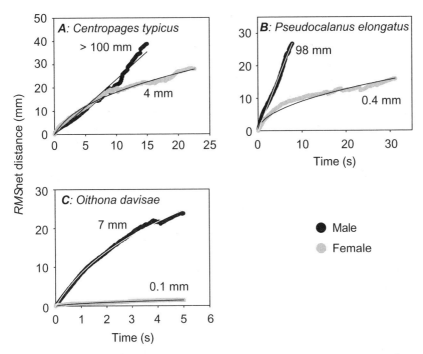

Fig. 4.16. Analyses of male and female motility patterns in three species of pelagic copepods that use pheromone signaling in mate finding. For each species and sex the root-mean-square distance (two dimensions) traveled as a function of time is shown. Taylor's equation has been fitted to the data, and the so-estimated motility length scales have been shown. Motility length scales of males are consistently about two orders of magnitude longer than those of the females.

range of its predators. This would optimize the trade-off between prey and predator encounter rates. Such an intermediate run length would restrict encounters with predators and yet have reasonably high encounter rates with prey. There is evidence that, over a large size range of planktonic organisms, motility patterns have such intermediate motility length scales (Visser and Kiørboe 2006).

The motility pattern of an organism may change with conditions. For example, starving organisms may adopt a more ballistic motility pattern than well-fed ones because it may more efficiently bring them to areas (patches) with better feeding conditions (for an example, see Menden-Deuer and Grünbaum 2006). This behavior, of course, would pose an elevated predation risk to the searcher, which therefore has to balance the trade-offs between gains and risks. Motility patterns may also differ between males and females because the two sexes may have different roles in mate finding. We have previously seen how female copepods produce pheromone signals that the males have to find. You may recall that the extension of the female signal is insensitive to her swimming velocity and pattern (section 3.9). A female can therefore move "predator-safe" and swim along a convoluted path with run lengths just long enough to exceed the sensory distance to her phytoplankton prey but much smaller than the sensory range of her predators. The male, in contrast, has to search for female signals of much larger spatial extent, and he should therefore adopt a much more directionally persistent swimming pattern with a longer motility length scale. Such a male-female difference in motility pattern appears to occur consistently among pelagic copepods with pheromone mate signaling (fig. 4.16). The male behavior implies a much larger risk of encountering a predator than that of the female. The males run this elevated risk because they gain more than they lose in terms of numbers of female encounters during their adult lives (which is essentially all that matters). The difference in swimming pattern and predator risk between sexes can best be explained by "asymmetric" interests in mate encounters between the sexes as a result of sperm competition (Kokko and Wong 2007) and may itself explain why sex ratios in copepod field populations typically are very female biased (Kiørboe 2006).

Chapter Five

HYDROMECHANICAL SIGNALS
IN THE PLANKTON

MANY PLANKTONIC organisms can perceive events in their environment, for example, an approaching predator, a passing prey, or a potential mate, and respond accordingly, i.e., escape, attack, or mate. Thus, they are able not only to perceive other plankters in their environment but also to identify and locate them. Mistakes can be critical! Few, if any, planktonic invertebrates have functional vision, and remote perception thus must be by chemical or hydromechanical senses. We have already talked some about chemical cues. This chapter examines hydromechanical signals in the plankton, and we consider in particular remote perception of predator and prey. Any organism moving through the water, or generating feeding currents, will generate a disturbance in the fluid that can be perceived. We shall examine first how signals are perceived and then how they are generated. Combining an understanding of the two allows us to estimate perception distances.

Visser (2001) provides an excellent and systematic overview of hydrodynamic signals in the plankton, but his account goes well beyond what is needed here. The following is mainly based on Kiørboe and Visser (1999). Their description has since been improved and corrected in many details (see review by Jiang and Osborn 2004), but the general principles of signal generation and perception laid out in the more accessible Kiørboe and Visser paper still apply (with minor corrections).

5.1 COPEPOD SENSORY BIOLOGY

In discussing fluid signals, we will for the sake of clarity be focusing on copepods. Many of the principles, or variants of the principles, also apply to other plankters. Copepods perceive hydrodynamic signals by means of setae that sit all over the body, but particularly on the distal parts, the antennules and the telson (fig. 5.1).

The setae are mechanoreceptors. If a seta is moved, it may elicit a neurophysiological response. Neurophysiological investigations have shown that the setae are velocity sensors rather than displacement sensors and

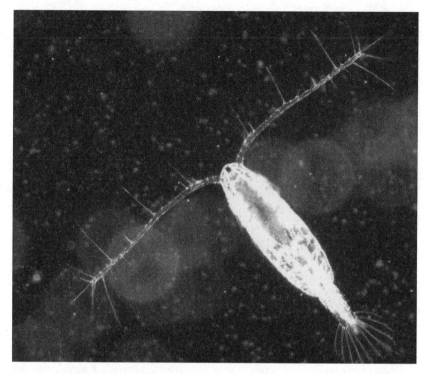

Fig. 5.1. Planktonic copepods, here *Acartia tonsa*, are well equipped with setae, sensory hairs that are placed mainly on the antennules and on the telson ("tail"). The setae are mechanoreceptors that perceive hydromechanical disturbances in the ambient fluid (Strickler and Bal 1973).

that the lower velocity threshold for a neurophysiological response is about 10–20 μm s^{-1} (Yen et al. 1992). The setae will bend only if there is a velocity difference between the copepod and the ambient water. Because planktonic organisms, by definition, are embedded in the general fluid flow, they cannot perceive flow per se, and flow or flow velocity, thus do not provide cues. A copepod following the general flow does not experience a velocity difference; hence, the setae will not bend. Hydrodynamic signals thus have to do with velocity gradients. All parts of a rigid prey copepod in a velocity gradient cannot travel with the ambient fluid: if the central part follows the fluid, the distal parts will travel at a velocity smaller or larger than the ambient fluid, and this may cause the setae to bend and thus potentially elicit, for example, an escape response (fig. 5.2). We will assume that this velocity difference is the signal perceived by the copepod and call the velocity difference the signal strength. For

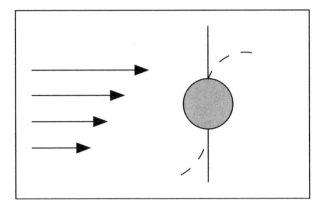

Fig. 5.2. Spherical copepod in a velocity gradient. Because all parts of the rigid copepod cannot travel with the same velocity as the fluid, there will be a velocity difference between the copepod and the ambient fluid, and extending setae will bend.

a perfect, rigid sphere the signal strength (*S*) is simply the velocity gradient multiplied by the radius of the sphere (we ignore boundary-layer effects). This, of course, assumes that the flow field is "large" relative to the copepod, for example, as generated by a large predator approaching the small copepod. If, on the other hand, a small prey swims past a predatory copepod, then the fluid disturbance generated is "small" or local, and will not embed the large predatory copepod. In this case, the velocity difference between the copepod and the ambient water is simply equal to the fluid velocity generated by the small prey.

5.2 DECOMPOSITION OF A FLUID DISTURBANCE: DEFORMATION AND VORTICITY

Before we continue our examination of signal perception, we need to introduce and define some terms and phenomena. For any fluid disturbance, the motion of a fluid element can be decomposed into three parts: translation, rotation, and deformation (fig. 5.3). These three elements can be quantified by, respectively, the fluid velocity, the vorticity, and the deformation rate (=rate of strain). The latter two components are related to velocity gradients (i.e., are zero in the absence of a velocity gradient), and quantitatively, the sum of the rotation and deformation equals the velocity gradient.

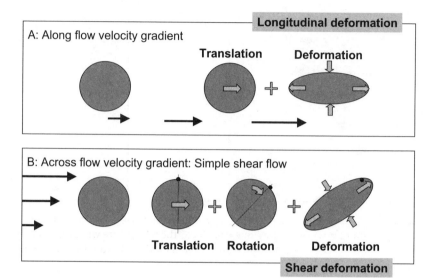

Fig. 5.3. Decomposition of a fluid disturbance. Velocity gradients can either be in the direction of the flow (a) or perpendicular to the flow (b), or a combination thereof. One can distinguish between stretching and shear deformation accordingly. Modified from Kiørboe and Visser (1999).

Although translation or fluid velocity is a well-known property, vorticity and deformation are perhaps less familiar. Vorticity describes the rotational component of the flow and, by definition, is two times the rotation rate. It has dimensions of (radians) per unit time. In a flow where the velocity gradient is only in the direction of the flow, there is no rotation, and the vorticity is zero (fig. 5.3A). In a flow where the velocity gradient is only in the direction perpendicular to the flow direction, vorticity equals the velocity gradient (fig. 5.3B).

To understand what deformation is, imagine the sphere in figure 5.3 as a fluid-filled balloon. In the presence of a velocity gradient the balloon will change shape—it will deform. Because the fluid inside the balloon is incompressible, its volume will remain unchanged, but the balloon will be stretched in one direction and compressed in another. One can quantify the deformation rate along any reference axis as the rate at which the two points on the periphery of the balloon that are intersected by the reference axis move away from or toward one another. This rate is expressed relative to the diameter of the sphere; hence, deformation also has dimensions of inverse time. In the following, we consider the "maximum deformation rate" as the deformation rate along the reference axis

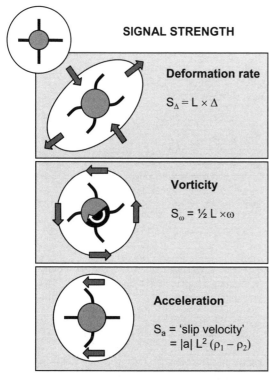

Fig. 5.4. Signal strength. Schematic of spherical copepod with extending setae in different flow environments. The different flow components give rise to different bending patterns of the setae and to velocity differences between the copepod and the ambient fluid, which result in signals (S) of different magnitudes. Modified from Kiørboe and Visser (1999).

that yields the highest rate. We shall simply refer to this as the deformation rate.

5.3 SIGNAL STRENGTH: PREY PERCEIVING PREDATOR

We can now examine which of the flow components of a fluid disturbance may cause a velocity difference between the copepod and the ambient water, and we can more explicitly define the signal strength (fig. 5.4). We have already made the point that the advective component causes no velocity difference for a small prey copepod embedded in the flow generated by a larger moving predator and hence produces no signal. Fluid deformation, on the other hand, causes a velocity difference and will make setae bend, as

TABLE 5.1
Signal Strength for Various Flow Components

Flow component	Signal strength	Equation number
Deformation	$a\Delta$	5.1
Vorticity	$1/2a\omega_h$	5.2
Acceleration	$|\mathbf{a}| \, a^2 (\rho_1-\rho_2) / 9\eta$	5.3
Translation small prey	u	5.4

a is the radius of the perceiving copepod, Δ is the deformation rate, ω_h is the horizontal component of the vorticity, $|\mathbf{a}|$ is the acceleration of the fluid, and η is the dynamic viscosity. For a large predator perceiving a small swimming prey, the signal strength is simply the flow velocity generated by the prey. Equations are taken from Kiørboe and Visser (1999).

depicted in figure 5.4. The signal strength is simply the deformation rate Δ multiplied by the radius of the (as usual spherical) copepod (table 5.1).

Vorticity may also cause a velocity difference between the copepod and the ambient water depending on the properties of the prey. If the center of mass of the copepod is identical to its geometric center, then the copepod will just rotate with the fluid, and there will be no velocity difference, hence no signal. However, if the center of mass is offset from the geometric center, such as if the copepod is "bottom heavy," then the copepod will tend toward a permanent position in a rotating fluid, and there will be a velocity difference. The signal strength caused by fluid rotation is a simple function of vorticity and radius (eq. 5.2 in table 5.1).

There is a final characteristic of fluid flow that we have not yet considered, namely flow acceleration. Whenever there are velocity gradients, there is acceleration. We have all experienced being pressed backward in an accelerating car or, more notably, an accelerating jet plane. This happens because we are heavier than the ambient medium. Because the typical copepod is slightly more dense than the ambient water, it will experience a similar "slip velocity" in an accelerating flow, i.e., a signal. The magnitude of this slip velocity is a function of the density difference, the acceleration, and the radius of the copepod, and the quantitative relationship is similar to Stokes' law (the law predicting the sinking velocity of a sphere) (eq. 5.4, table 5.1).

5.4 SIGNAL STRENGTH: PREDATOR PERCEIVING PREY

A large predator is not embedded in the flow generated by a small, swimming prey, and the signal strength in this situation is simply the flow velocity (which attenuates with the distance to the prey—see below) (eq. 5.4, table 5.1).

5.5 To What Flow Components Does a Copepod Respond?

How can a copepod interpret a hydrodynamic signal and respond adequately? It is equipped with an array of setae that are oriented in all directions, and different setae have different bending characteristics and neurophysiological sensitivities (e.g., Fields et al. 2002). Together, this may in principle allow the copepod to paint quite a detailed three-dimensional picture of the moving object. For example, the fluid signal generated by a small prey may affect only a few nearby setae or affect the different setae with very different intensities, whereas the large flow field generated by a large predator affects all setae. Thus, a "local flow field" would suggest attack, whereas a "global flow field" would suggest escape. This has been demonstrated very elegantly in a chaetognath that also perceives prey and predators by means of mechanosensory setae: stimulation of a few setae elicits an attack response, but simultaneous stimulation of many setae elicits an escape response (Nishii 1998). It probably works much the same way in copepods.

A copepod may in principle also be able to distinguish between signals by the different flow components of the larger flow field generated by an approaching predator because the different components result in different setal bending patterns (fig. 5.4). This may further help the copepod to characterize the approaching object because the exact composition of the signal will depend on the velocity and distance of the object. However, one cannot from pure reasoning decide to which, if any, of the flow components a copepod responds. This can be decided only from experiments. Although any natural hydrodynamic signal contains deformation, vorticity, and acceleration components simultaneously, one may experimentally generate situations in which the components have been separated, thus allowing one to examine the responsiveness to the different signal components. Such an experimental setup has been described in figure 5.5. For example, the flow generated in front of a suction pipette is characterized by (longitudinal) deformation and acceleration, but there is no rotational component (zero vorticity). Copepods exposed to such a flow will all respond by attempting to escape: first the copepod is entrained in the flow, but at a certain distance to the pipette tip, which depends on the flow velocity, it will perform powerful escape jumps at velocities of > 100 body lengths s^{-1}. This has been demonstrated repeatedly for many species of copepods (e.g., Singarajah 1969, Fields and Yen 1997, Kiørboe et al. 1999). From such an experiment one can conclude that the copepod responds either to deformation or to acceleration or to both. In a closed

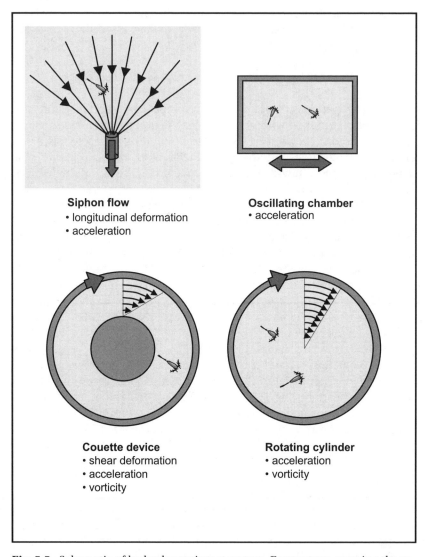

Siphon flow
- longitudinal deformation
- acceleration

Oscillating chamber
- acceleration

Couette device
- shear deformation
- acceleration
- vorticity

Rotating cylinder
- acceleration
- vorticity

Fig. 5.5. Schematic of hydrodynamic test system. Four setups examine the responsiveness of zooplankton to the different components of a hydrodynamic signal. Modified from Kiørboe et al. (1999).

chamber that oscillates, there is only acceleration of the flow—no deformation, no vorticity. *A. tonsa* copepods do not respond to even very high acceleration intensities in such a chamber. Thus, we may conclude that *A. tonsa* responds to deformation but not to acceleration. One may similarly expose copepods to flows generated in rotating chambers. In a Couette device—a cylindrical chamber rotating around a solid, nonrotating cylinder—the flow in the annular gap between the chamber wall and the inner cylinder is characterized by shear deformation, vorticity, and acceleration; the flow in a rotating cylinder has acceleration and vorticity but no deformation. Only the flow in the Couette device elicits escape responses in *A. tonsa*. Thus, we conclude that they do not respond to vorticity. The overall conclusion is that escape responses in *A. tonsa*, and probably most other copepods, are elicited only by signals generated by deformation. A similar conclusion has been arrived at for unicellular pelagic protists (Jakobsen 2001). Protists do not have mechanosensory setae, but presumably the entire cell functions as a deformation sensor, similar to the deforming balloon considered above.

5.6 SENSITIVITY TO HYDRODYNAMIC SIGNALS

Experiments of the kind described in figure 5.5, in particular the siphon experiment, also allow one to estimate the critical deformation rate required to elicit an escape response. In the siphon flow, the deformation rate depends on the flow rate and declines with the distance to the tip of the pipette in a known manner. Thus, if we note the distance at which an escape response is initiated, then we have an estimate of the critical deformation rate, Δ^*. Critical deformation rates estimated for a variety of copepods and other zooplankters are in the range 1–10 s^{-1} (data compilation in Kiørboe et al. 1999).

Thus, this is the order of magnitude of deformation rates required to elicit escape responses. Because the signal strength generated by deformation is given by $S = a\Delta$ (eq. 5.1), we can also estimate a critical signal strength (velocity difference) as $S^* = a\Delta^*$. Estimates of S^* vary much more among zooplankters than estimates of Δ^* and, in addition, are typically orders of magnitude higher than the signal strength required to elicit a neurophysiological response. We shall later discuss why this may be.

5.7 PREDATOR AND PREY REACTION DISTANCES: GENERATION OF A HYDRODYNAMIC SIGNAL

Having some insight in the hydromechanical sensitivities of copepods and other plankters, we may now ask what the fluid disturbances produced by a moving plankter are. In particular, we want to know how flow velocity (prey detection) and deformation rate (predator detection) attenuate with the distance to the object that generates it. Equipped with such knowledge, we will able to predict detection or response distances (R).

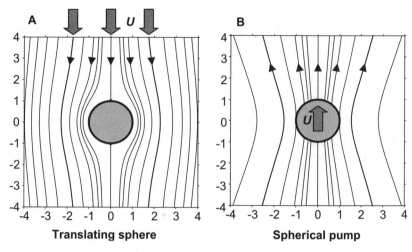

Fig. 5.6. Stokes' flow: streamlines around a sphere for the translating sphere and the spherical pump solutions. Modified from Kiørboe and Visser (1999).

Several authors have mapped flow fields around swimming and filter-feeding plankton organisms (e.g., van Duren et al. 1998, Doall et al. 2002, Catton et al. 2007), and this allows computation of both flow velocities and deformation rates as functions of distance to the organism. Hence, the goal is accomplished, but only in such specific cases. We shall here attempt a more generic approach by presenting simple models of the flow field generated by moving and filter-feeding plankters. We shall consider two models, one that describes the plankter as a translating sphere under the action of a body force at low Re and one in which the moving plankter is considered a self-propelled body and is approximated by a "force dipole." The former model is the Stokes' flow solution that we considered above (sections 3.3 and 4.2, eqs. 4.1 and 4.2). Both models are, of course, extreme simplifications of the real world, and both have their pros and cons. The translating-sphere model assumes a "body force" such as, for example, a sphere sinking under gravity. In reality, most organisms move by means of swimming appendages, not a body force. The dipole model takes this into account in a simple way by considering the swimming organism as two equal but oppositely directed point forces (thrust and drag) working on the fluid. This model thus disregards the effect of the presence of the body on the flow field (except for the drag force). For a swimming plankter, the translating-sphere/body force model is probably most accurate near the organism, whereas the dipole model is better in the far field. We shall therefore apply the two different models to, respectively, prey perceiving

Table 5.2
Reaction Distances to Predator and Prey

	Reaction Distance	Equation Number
Prey perceiving predator	$R(\theta = 0^0) = a\sqrt{\dfrac{3U}{4a\Delta^*}}\left\{1 + \sqrt{1 - \dfrac{8a\Delta^*}{3U}}\right\}$	5.6
Predator perceiving prey	$R = a\sqrt{\dfrac{3U}{2S^*}}\left|3\cos^2\theta\right|$	5.8

a large predator (near field) and predator perceiving small prey (far field).

We are already familiar with Stokes' creeping flow around a sphere (eqs. 4.1 and 4.2), which we shall use to examine predator detection distances. However, there are two solutions to the Stokes' flow around a sphere, one where the flow is around a sphere fixed in space (fig. 5.6A) and another where the flow is around a moving sphere (fig. 5.6B). The two solutions are in fact identical and related by a coordinate transformation, i.e., coordinates fixed on the sphere (fig. 5.6A) or fixed in space (fig. 5.6B). The two different (but similar) solutions in figure 5.6 may be interpreted as the flow around a moving plankter (fig. 5.6A) or the flow due to a plankter that generates a feeding current (fig. 5.6B). Thus, one model encapsulates two very general plankton feeding types, a cruising predator and a suspension feeder.

Because a predator typically approaches a prey head on (or the prey arrives from the front of the predator in the feeding current), and because the prey may detect the signal generated by fluid deformation, what we really want to know is how deformation rate varies directly in front of the predator. This can be estimated as the derivative of the flow velocity in the radial direction (eq. 4.1) at $\theta = 0°$, thus

$$\Delta(\theta = 0^0) = \frac{\partial u_r}{\partial r}(\theta = 0^0) = 3Ua(r^2 - a^2)/(2r^4) \tag{5.5}$$

The deformation rate is zero at the "surface" of the predator, increases to a peak value, and then attenuates rapidly (fig. 5.7a).

We are now able to estimate the distance at which a prey will perceive an approaching predator (or realize that it has been entrained in a feeding current), R, if we know the threshold deformation rate required to elicit an escape response, Δ^*. We just insert Δ^* in equation 5.5 and solve for r. The result is given as equation 5.6 in table 5.2, and the solution has

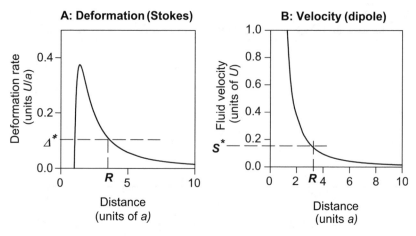

Fig. 5.7. Fluid signals generated by moving predator and prey. The variation with distance in fluid deformation (A) or fluid velocity (B) directly in front of a translating sphere (or spherical pump) and a force dipole, respectively. A prey organism that responds to a critical deformation rate Δ^* will perceive an approaching predator at distance R (A). Similarly, a predator that reacts to a critical local flow velocity S^* will perceive an approaching prey at distance R (B).

also been illustrated graphically in figure 5.7A. (From fig. 5.7A one can see that there are two solutions; obviously, only the solution further away is of interest.)

Despite its simplicity, the model is quite efficient in predicting the distances at which copepod prey react to approaching predators or when entrained in predator feeding currents for a variety of predator-prey systems (fig. 5.8).

The dipole model applied to describe the fluid disturbance generated by a small swimming prey similarly predicts a distance at which a predator reacts to the prey (Svensen and Kiørboe 2000, Visser 2001). According to the dipole model, the flow velocity (u) generated by a self-propelled plankter of radius a and swimming with velocity U as a function of distance (r) and direction (θ) is:

$$u = \frac{3Ua^2}{2r^2}(1 - 3\cos^2\theta) \tag{5.7}$$

Putting $u=S^*$, the threshold sensitivity, and solving for r yields the reaction distance (eq. 5.8, table 5.2). Again we can provide a graphic solution (fig. 5.7B). The only empirical test of this model that I am aware of is that of Svensen and Kiørboe (2000), who used it to predict the reaction distance of the ambush-feeding copepod *Oithona similis* to the swimming

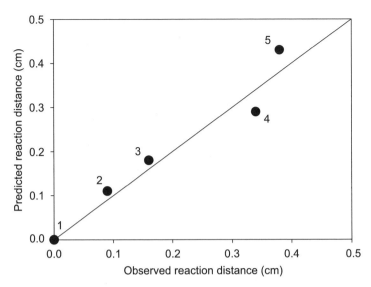

Fig. 5.8. Observed and predicted (eq. 5.6) prey reaction distances in a number of predator-prey interactions. 1, Small fish versus a copepod (Stickleback–*Temora*); 2, adult copepod versus copepod nauplii (*Centropages–Acartia* nauplii); 3, adult copepod versus copepod nauplii (*Temora–Acartia* nauplii); 4, small fish versus a copepod (Stickleback–*Eurytemora*); 5, larval fish versus copepod nauplii (cod–*Acartia* nauplii). Data are from Viitasalo et al. (1998). (1,4), Titelman (2001) (2), Yen and Fields (1992) (3), and Jens Rasmussen, Danish Institute for Fisheries Research (unpublished, 5).

dinoflagellate *Gymnodinium dominans*. The observed reaction distance averaged 140 μm, whereas predicted reaction distance varied between 110 and 160 μm depending on the approach angle (θ).

5.8 ATTACK OR FLEE—THE DILEMMA OF A PARASITIC COPEPOD

Many marine copepods are parasitic, and most of them have a planktonic infective stage. To these stages an approaching fish may be either a predator, a potential host, or both. Chemosensory information is generally not useful to decide the presence or identity of an approaching object because the time scale of molecular diffusion is too long to provide sufficiently early warning. In an interesting study, Heuch et al. (2007) compared the responses of a holoplanktonic copepod (*Acartia*) and the infective stages of a parasitic copepod (salmon louse, *Lepeophtheirus salmonis*) to an approaching fish mimic. The holoplanktonic copepod responded in most cases by powerful escape jumps away from the fish, as one would expect. The parasitic copepod, in contrast, responded in most cases to the approaching fish mimic by swimming toward it in a characteristic, continuous, spiraling pattern. In many cases the parasitic copepod would initially swim away from the mimic and then turn around to approach the fish mimic from its side as it passed by. The latter response appears to have particularly high survival value because it reduces the risk that the copepod is eaten by the fish and at the same time enhances its chances of attacking it. The responses of the copepods were independent of light, and the fish mimic provided no chemical signals; hence, only hydromechanical signals were received by the copepods. The fluid deformation rates generated by the fish mimic at the distance at which the two copepods responded were similar to what one can measure in siphon experiments as described above (>0.5 s^{-1}). Thus, the same hydromechanical information leads to very different and adequate responses in two species of copepods.

5.9 Maximal Signals, Optimal Sensitivity, and the Role of Turbulence

We noted above that the fluid deformation generated by a predator has a maximum at some distance in front of the predator (fig. 5.7A). From equation 5.5, it can be shown that the magnitude of this maximum deformation rate is given by

$$\Delta_{max} = \frac{3U}{8a} \tag{5.9}$$

As an example, the small copepod *Paracalanus parvus* has a maximum feeding current velocity of $U=0.4$ cm s^{-1} and a radius of about $a=0.015$ cm (Tiselius and Jonsson 1990); hence, $\Delta_{max}=10$ s^{-1}. One can do similar calculations for a swimming fish larva (Kiørboe et al. 1999) and for a ciliate feeding current (Jakobsen 2002), and one will come up with very similar estimates of Δ_{max}. In fact, the maximum deformation rate produced by almost any small planktonic predator is on the order of 10 s^{-1}. The reason is that Δ_{max} is proportional to the ratio of swimming (or feeding current) velocity to body size (a), and this ratio is almost constant because swimming velocity scales with body size (approximately). Thus, typical maximum deformation rates of small planktonic predators are of the order 10 s^{-1}. This, in turn, implies that planktonic prey organisms should have critical deformation sensitivities less than this, which in fact appears to be the case (fig. 5.9).

Of course, the lower the critical deformation sensitivity of a prey organism, the further away it will be able to perceive an approaching predator, and the better is its chance of escaping. However, apparently

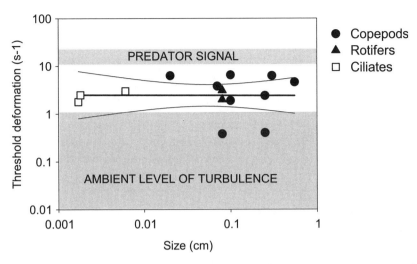

Fig. 5.9. Threshold deformation rate required to elicit escape responses in various zooplankters as a function of organism size. The upper gray bar shows the typical maximum deformation rate produced by a small planktonic predator, and the lower gray area the range of deformation rates caused by typical and maximum levels of ambient turbulence in the ocean surface layer. Critical deformation rates fall in the window between the two, i.e., low enough to perceive predators remotely and high enough to avoid inordinate escape reaction to turbulence (noise). Data for copepods and rotifers were compiled by Kiørboe et al. (1999); data for ciliates are from Jakobsen (2001, 2002).

copepods, at least, appear to be far from utilizing their neurophysiologic potential. They respond only to deformation signals that are much stronger than the weakest signals that they can perceive. A likely explanation of this apparent paradox is that the copepod must avoid noise-elicited inordinate escapes because this would disturb its feeding (Fields and Yen 1997). Turbulence causes the water to deform, and one would predict copepods inhabiting deeper, quiescent waters to have lower response thresholds than those inhabiting the more turbulent surface waters (Fields and Yen 1997), but this idea has not yet been adequately tested, partly because most observations are for coastal zooplankters. Realistic turbulent intensities in coastal water and in the upper ocean produce fluid deformation rates of up to about $1 \, s^{-1}$. We would thus expect realized deformation sensitivities to fall in the "window" between the minimum requirement for predator detection ($10 \, s^{-1}$) and the turbulence-generated fluid deformation rates ($\sim 1 \, s^{-1}$). Estimates of critical deformation rates based on observations from the literature support this understanding (fig. 5.9).

5.10 THE EVOLUTIONARY ARMS RACE

Copepods make up about 80% of the mesozooplankton in the oceans and are, thus, in an evolutionary sense a very successful group (Verity and Smetacek 1996). One striking feature of almost all pelagic copepods is that they look very much the same (a challenge to taxonomists and a nightmare to ecologists; fig. 5.10): a torpedo-shaped body that hydrodynamically allows for high swimming velocities and antennules sticking out in the water densely equipped with sensors. Copepods are sensory escape machines! And in fact, escape velocities of copepods are impressive: several hundreds of body lengths per second. Because of the hydrodynamic signal that cruising and suspension-feeding plankters generate, and because of the ability of copepods (and many other plankters) to detect such signals, one would expect simple suspension- or cruise-feeding strategies to be inefficient in capturing copepods in particular and motile prey in general. In the course of the evolutionary arms race, however, many predators have partly circumvented prey defense mechanisms in various ways, so the "bugs" are still not safe. Yen and Strickler (1996), for example, describe how naupliar prey entrained in the feeding current of a predatory copepod make strong escape jumps elicited by the feeding current. However, the escape jumps signal the presence and exact location of the prey to the predator, which consequently attacks the prey.

Another example is provided by fish larvae feeding on copepods. Many fish larvae are cruising predators that locate their prey visually, and they are able to see the prey further away than the prey can detect the approaching fish larva. Immediately on prey detection, the fish larva decelerates, assumes an S-shaped attack posture, and then approaches the prey very slowly (Munk and Kiørboe 1985), thus allowing the predator to come to striking distance unnoticed. Because there is a maximum deformation rate produced by the approaching larval fish that is a function of the approach velocity, as given by equation 5.9, it follows that there is a critical swimming velocity, $U_{Cr} = 8a\Delta^*/3$, below which the prey will not notice the approaching predator. Thus, the approach speed has to be less than U_{Cr}. There is a trade-off here, however, because the slower the approach, the greater the risk (or chance) that the prey will move on or that turbulence will advect the prey out of the predator's dining sphere (cf. section 4.9). Therefore, the optimum approach speed is just less than U_{Cr}. Viitasalo et al. (1998) showed that those individuals of the small fish the three-spined stickleback that successfully captured copepod prey were indeed those with approach velocities just below U_{Cr}. Thus, there are ways to trick copepods.

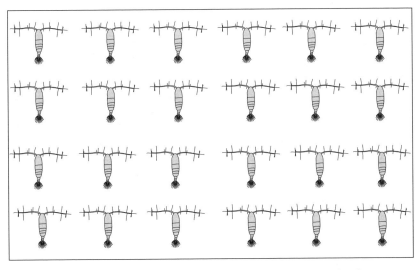

Fig. 5.10. Schematic illustrating the morphological diversity of pelagic copepods. The morphologies of most pelagic copepods are all very similar in that they have well-developed sensory machineries and torpedo-shaped bodies and are, thus, efficient in detecting and escaping predators. This suggests that predation has been a very strong selective force in shaping the morphology of this successful group of zooplankton (Verity and Smetacek 1996).

This "watery arms race" applies, of course, not only to copepod predator-prey interactions but also to other zooplankters and to phytoplankton (Smetacek 2001). There are many examples of predator protection mechanisms among phytoplankton and other protists: colony formation brings small cells into a size range where predation pressure is less (chapter 8), and colonies of the often dominating *Phaeocystis* are further enclosed in a tough skin that provides protection (Hamm et al. 1999) much the same way that the very strong frustules in diatoms provide protection from small predators (Hamm et al. 2003). Siliceous spines on diatoms may similarly make them difficult to ingest for some predators; many protists can perceive hydrodynamic signals and respond by escape jumps such as we saw with the copepods (Jakobsen 2002); and toxin production is common among some groups of phytoplankters and may reduce their susceptibility to predation (Teegarden 1999). Toxin production in some dinoflagellates may even be triggered by the presence of grazers (Selander et al. 2006). In fact, in the absence of predation, all phytoplankters would be very small (chapter 8), and the diversity in the morphology and ecology of phytoplankton may to a large extent be interpreted as adaptations to predator avoidance rather than as adaptations to light and nutrient acquisition (Smetacek 2001). However, in all cases the protection is incomplete, and many planktonic predators have "learned" though evolution to overcome the protection mechanisms. For example, copepods may clip off spines from diatoms before ingesting them (R. Strickler, personal communication), populations of copepods that are used to experiencing toxic algal blooms may develop immunity to the toxin that inexperienced populations are lacking (Colin and Dam 2003, 2004), and escaping protists may elicit attacks in protist predators similar to what was reported for copepods above (Jakobsen et al. 2006). Coevolution through such arms races may be the dominant force guiding the evolution of plankton (Smetacek 2001).

Chapter Six

ZOOPLANKTON FEEDING RATES
AND BIOENERGETICS

ENCOUNTERS between predator and prey may lead to the former eating the latter. Consumption, of course, is a prerequisite for organism growth and reproduction and, in turn, constitutes the input to population processes, such as population growth and mortality. Encountered prey is not necessarily consumed, however. The prey has to be pursued, captured, and handled before it is ingested, but in any of these steps the prey may be lost. Also, any particular prey item may be selected against, for example, because the predator is satiated or the prey unsuitable. Ingested prey is utilized for growth and reproduction, but only after metabolic expenses have been covered. This chapter explores the components of planktonic predation and examines how ingested food is converted to growth and reproduction. It thus draws on preceding chapters (encounter rates) and provides input to subsequent chapters (population processes).

6.1 FUNCTIONAL RESPONSE IN INGESTION RATE
TO PREY CONCENTRATION

The dependence of prey ingestion rate on prey concentration is termed the *functional response*. Although prey encounter rate increases in proportion to prey density, the functional response in ingestion rate typically saturates at high prey densities. Based on the work of Holling (1959a) one can distinguish three types of functional responses: a linear increase, a decelerating increase, and an S-shaped increase in ingestion rate with prey concentration (fig. 6.1). One can also plot the corresponding responses in clearance rate to prey concentration, which often allows a better separation of the responses (e.g., from experimental observations). Examples from the plankton world of observed functional responses are numerous, and some have been given in figure 6.2.

The most basic functional response is a type II response. Mathematically this response is described by the so-called "disk equation" of Holling (1959b). Holling explored the nature of the functional response by examining an artificial predator–prey situation by allowing his

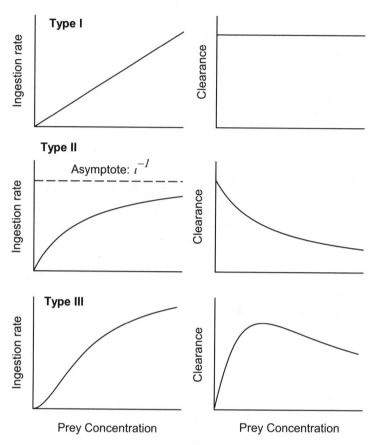

Fig. 6.1. Holling's three functional response types, plotted both as ingestion rate and as clearance rate as functions of prey concentration (1959a).

blindfolded assistant to act as a predator searching for prey, which were sandpaper disks that were spread on a table; hence the unexpected name of the equation. Its derivation is straightforward. From equation 1.1b we know that the prey encounter rate is βC_{Prey}. Therefore, during search time T_S, the predator will encounter N prey:

$$N = \beta C_{\text{prey}} T_s \tag{6.1}$$

Assume now that it takes a certain time, ι, to handle each prey. This handling may include the time to pursue, capture, ingest, and digest the prey. The time required to handle the N encountered prey is

$$T_h = \iota N \tag{6.2}$$

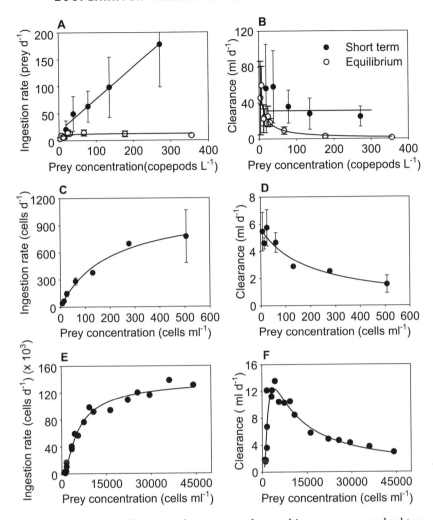

Fig. 6.2. Examples of functional responses observed in common zooplankton organisms, each representing one of Holling's three response types. Data for each species are plotted both as ingestion rate and as clearance as a function of prey concentration: the jellyfish *Sarsia tubulosa* (see Fig. 6.3B) feeding on copepods (*Acartia tonsa*) (A, B) (Hansson and Kiørboe 2006a), the copepod *Oithona davisae* (Fig. 6.3A) feeding on dinoflagellates (*Oxhyrris marina*) (C, D) (Saiz et al. 2003), and the copepod *Acartia tonsa* feeding on small flagellates (*Rhodomonas salina*) (E, F) (Kiørboe et al. 1985). There are two sets of curves fitted to the data for *Sarsia*, one for short-term feeding rate subsequent to starvation (Holling type I) and one for long-term feeding rate. The curves fitted to the data for *Oithona davisae* are Holling's disk equation, and the parameter estimates are the encounter-rate kernel, $\beta = 5.6\,\text{ml d}^{-1}$, and the handling time, $\tau = 1.3\,\text{min prey}^{-1}$. The latter implies a maximum ingestion rate of $\tau^{-1} = 1100$ prey cells d^{-1}.

and the total time required to find *and* handle the N prey is

$$T_T = T_s + T_h \tag{6.3}$$

Combining equations 6.1–3 and noting that the ingestion rate is $i = N/T_T$ yields

$$i = N/T_T = \frac{\beta C_{prey}}{1 + \beta \iota C_{prey}} \tag{6.4}$$

The disk equation is identical to the Michaelis–Menten equation used to describe enzyme kinetics. The formulation and derivation as well as the interpretation of the parameters, however, are more relevant to a predator-prey situation. One parameter from the Michaelis–Menten formulation may be useful, the half-saturation constant (K), which relates to the parameters of the disk equation as $K = (\beta \iota)^{-1}$. From the disk equation it follows that, at low prey concentration, ingestion is limited by prey encounters, and the ingestion rate tends toward the prey encounter rate, βC_{prey}; at high prey concentrations, handling limits ingestions, and the ingestion rate tends toward the inverse of the prey–handling time, ι^{-1}. In figure 6.2C, Holling's disk equation has been fitted to the observed functional responses in a small zooplankter.

6.2 EXAMPLE: THE FUNCTIONAL RESPONSE IN *OITHONA DAVISAE*

There are two parameters in the disk equation, β and ι. Knowing them, we know the functional response. The two parameters may either be estimated from curve fitting, as in figure 6.2C, or they may be estimated independently from insights into prey perception, encounter, pursuit, capture, and digestion. The latter—mechanistic—approach is much preferred to "blind" curve fitting because it allows much safer extrapolation of the observations to scenarios not examined experimentally, e.g., to other prey types or temperatures. As an example of direct estimation of the parameters, consider the tiny ambush-feeding copepod, *Oithona davisae* (fig 6.3A). This copepod has a typical Holling type II functional response in ingestion rate to the concentration of its prey, the flagellate *Oxhyrris marina* (fig. 6.2C). The females of *Oithona* rarely swim but remain motionless and, thus, sink very slowly through the water while scanning for motile prey. Moving prey are perceived hydrodynamically, as described in the previous chapter, and one can estimate prey reaction distances from size and swimming speed of the prey (eq. 5.8) and, hence, the encounter rate kernel, β (eq. 4.12 or a variant thereof, see Svensen and Kiørboe 2000). By this method, Saiz et al. (2003) estimated $\beta = 6$ ml d^{-1}

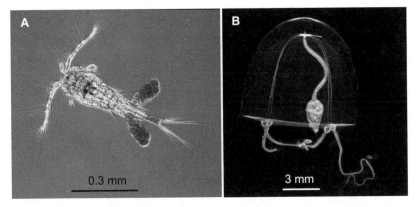

Fig. 6.3. The cyclopoid copepod *Oithona davisae* (A) (courtesy of Albert Calbet) and the hydrozoan jellyfish *Sarsia tubulosa* (B) (courtesy of Russ Hopcroft).

for adult females. This is very similar to the estimate derived from curve fitting (see legend of fig. 6.2C). We can also estimate the prey handling time. In *O. davisae*, perceived prey is attacked in a rapid jump of the copepod toward the prey, and the prey is grabbed and engulfed in less than a second, that is, very fast. Prey handling is therefore primarily made up of the time necessary for passage through the gut. At the experimental temperature (21°C), the gut turnover time is about 20 minutes in copepods (Dam and Peterson 1988). From the size of the prey (~2000 μm^3) and the volume of the gut (30,000 μm^3, assumed equal to the size of two fecal pellets), it takes about fifteen prey cells to fill the gut. Hence, $t = 20/15$ min prey^{-1} = 1.3 min prey^{-1}. Again, the independent and direct estimate of the parameter is similar to that estimated from curve fitting. Together the parameters estimated from an analysis of the component processes provide a good prediction of the functional response actually observed. The approach used here allows predictions of the functional responses to concentrations of other prey organisms as long as we know their size and swimming velocity.

6.3 OTHER FUNCTIONAL RESPONSES

There are two basic assumptions in the disk equation, namely that the handling time (t) and the encounter-rate kernel (β) are independent of prey concentration. Although these assumptions appear to be fulfilled in the *Oithona* example above (female feeding behavior is invariant with prey concentration), this is not always the case. For example, the gut throughput time may depend on the amount of food in the gut, typically decreasing with increasing gut fullness (Penry and Jumars 1986). This may lead to a functional response that in practice is indistinguishable from a type II response. Thus, an acceptable statistical fit of the disk equation to observations is not necessarily proof of the underlying mechanism, which is another argument that actually examining the mechanisms is a much stronger approach.

Another very likely deviation from the assumptions of the disk equation is that the feeding activity varies with the concentration of prey, making β dependent on the prey concentration. One response observed in several suspension-feeding copepods is that the generation of the feeding current from which prey is captured ceases or is reduced at low concentrations of phytoplankton prey (Price and Paffenhöfer 1986, Gill and Poulet 1988). Such a response would lead to a Holling type III functional response. This applies, for example, to suspension feeding in the copepod *Acartia tonsa,* where beating of the feeding appendages decreases with decreasing concentration of phytoplankton prey (Jonsson and Tiselius 1990), which explains the S-shaped functional response in ingestion rate to prey concentration observed in this species (fig. 6.2E) and many other copepods (Frost 1975, Mullin et al. 1975). Several other mechanisms—for example, prey switching (see below)—may give rise to a type III response, and it is not possible to write a general model for such a response. A type III response implies positively density-dependent prey mortality over a range of prey concentrations, i.e., increasing predation mortality with increasing prey density, and thus has the potential to act as a population regulation mechanism. This type of response, and the mechanisms causing it, have therefore attracted considerable interest among population ecologists and modelers (Murdoch and Oaten 1975, Steele 1977).

A type I response, where the ingestion rate is an ever-increasing function of the prey concentration, is, of course, not really feasible in the long run because any predator has a limited capacity to handle and process captured prey. However, something near a type I response may apply ephemerally when a predator initiates feeding. Consider, for example, a visual predator starting to feed at dawn or a predator living in an otherwise meager environment entering a dense patch of food. The latter may be exemplified by the small jellyfish *Sarsia tubulosa* (fig. 6.3B). *Sarsia* captures prey by means of the tentacles hanging from the umbrella, and captured prey are transferred to the mouth and subsequently ingested, very similar to the feeding process in many other medusae. The capture-rate capacity is very high, and the gut voluminous, resulting in an apparently ever-increasing ingestion rate with prey concentration. This allows efficient utilization of a patchy food source. However, eventually, ingestion rate becomes limited by the digestive capacity in the gut. Starving individuals incubated for short periods may therefore show a type I response but a type II response after longer incubation periods (fig. 6.2A,B).

6.4 THE COMPONENTS OF PREDATION: PREY SELECTION

The ingestion of a prey is the final result of a series of sequential events: encounter, pursuit, attack, capture, handling, and ingestion. The rates, durations, and efficiencies of each of these components of the predation process as well as their dependencies on food availability together define the functional response, as discussed above. However, even for a specific predator all of these parameters may be different for different prey. This implies that different prey are ingested at different rates and, hence, that feeding is selective. Prey selection in planktonic predators typically is mainly a passive process. Passive prey selection implies that prey selection in a mixed prey assemblage can be predicted from single-prey experiments.

Consider as an example again the little *Sarsia* jellyfish feeding on three types of zooplankton prey (fig. 6.4) (Hansson and Kiørboe 2006b). Encounter kernels can be estimated by the principles described in previous chapters (chapters 2–4), and they differ by a factor of three among the three small crustacean prey because of different motility characteristics and sizes: large and rapidly swimming prey are encountered more frequently than small, slowly moving ones. Encountered prey may be captured on the tentacles, subsequently attached to the mouth, and finally ingested. The efficiencies of each of these processes again vary among prey species, and the products of encounter kernels and sequential handling efficiencies result in very different prey specific clearance rates and, hence, selective feeding. The mechanics of the selection are in this case entirely passive and do not imply any active preference of the predator.

Prey selection in planktonic predators is primarily governed by prey size. It may be modified by size-independent chemical, behavioral, or other characteristics of the prey, but generally, size transcends such secondary selection mechanisms when prey selection is examined over a broad (size) range of planktonic predators. The reason is that encounter rates in general will increase with prey size because larger prey move faster and are detected at greater distances than small prey. Conversely, prey capture efficiency typically declines with prey size because larger prey escape more efficiently and are more difficult to handle than smaller ones. As a consequence, the typical prey size spectrum becomes

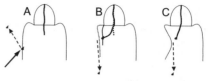

	Encounter kernel ml h⁻¹		Capture success		Attachment success		Ingestion success		Clearance
Cirriped nauplii	209	x	0.58	x	0.17	x	0.86	=	18
Cypris larvae	115	x	0.61	x	0.14	x	1.00	=	10
Copepod, *Acartia*	70	x	0.93	x	0.96	x	0.92	=	58

Fig. 6.4. Prey encounter and feeding rate in *Sarsia*. Small *Sarsia* medusae (Fig. 6.3B) encounter three small crustacean prey taxa at different rates, here quantified as encounter kernels. During the subsequent handling processes, many prey are lost, some types more than others, and the resulting clearance rates differ significantly among prey species. Modified from Hansson and Kiørboe (2006).

dome-shaped. An example of such size-dependent detection and capture efficiencies that combine into a dome-shaped prey size spectrum is provided by larval herring feeding on zooplankton (fig. 6.5), but similar prey size spectra are found much more generally among planktonic predators (fig. 6.6). Mechanical constraints—such as mouth size—imply that there is an optimum prey-to-predator size ratio defined on the basis of length that is characteristic within groups of organisms (fig. 6.6A). Optimum prey-to-predator size ratios therefore vary between groups but scatter around ~0.1 for planktonic predators (fig. 6.6B). In extreme cases, much larger or smaller prey are selected. One such fascinating example is provided by heterotrophic dinoflagellates that feed on prey that may be of the same size as or even larger than the predatory dinoflagellate itself (e.g., Hansen and Calado 1999, Naustvoll 2000). This is possible because many heterotrophic dinoflagellates do not engulf the entire prey cell but either suck out the prey cell contents by means of an extensible peduncle ("tube feeders") or digest prey cells externally by means of a pallium (Hansen and Calado 1999). At the other extreme, one finds examples of predators utilizing the "large-yet-small strategy" to increase their feeding rate, such as millimeter- to centimeter-sized pelagic tunicates that feed preferentially on bacteria and

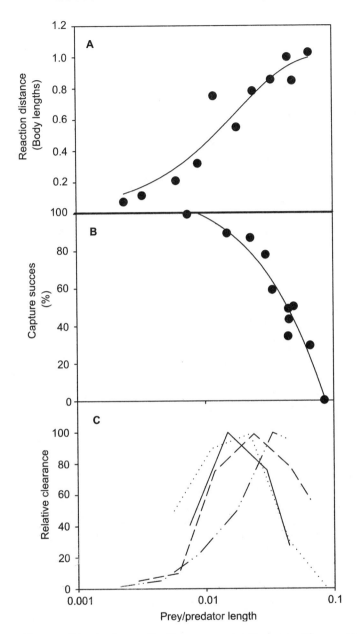

Fig. 6.5. Herring larvae feeding on zooplankton prey. The reaction distance to prey increases with prey size (A), whereas the capture success declines with prey size (B). As a result, the prey size spectrum (quantified by relative clearance rates) is dome-shaped (C). The different lines in panel C refer to larvae of slightly different sizes. Data from Munk (1992).

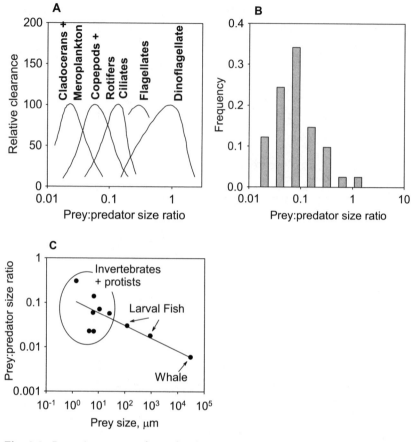

Fig. 6.6. Prey size spectra for individual groups of zooplankton predators (A) and frequency distribution of reported optimum prey:predator size ratios for 41 species of planktonic predators (B). In panel C the prey:predator size ratio as a function of prey size includes planktonic predators as well as predators feeding on plankton (baleen whales). A is redrawn from Hansen et al. (1994), and B and C are based on data in Hansen et al. (1994) and Munk (1992, 1997).

nanophytoplankton that are three to four orders of magnitude smaller than the predator, or voluminous jellyfish that feed on small microplankton prey. When the entire range of pelagic predators is considered, including baleen whales feeding on krill, there is a declining trend in the optimum prey:predator size ratio with increasing prey size (fig. 6.6C). The variable but constrained size ratio between predator and prey has implications for the structure of the pelagic food web that we shall return to later (chapter 8).

Prey selection may in some cases be based on criteria different from size alone, for example, the nutritional quality or the behavior of the prey. Such selection happens within the constraints posed by size. The *Sarsia* jellyfish considered above provides one such example of prey selection that is dependent on prey behavior rather than prey size alone. A similar example of behavioral-based prey selection is provided by the ctenophore *Mnemiopsis leidy* feeding on copepods (Waggett and Costello 1999). Differences in prey encounter rates between cruising and ambush-feeding nonmotile prey lead to preferential feeding on the former because capture efficiencies in this case were similar. Prey selection based on differences in the behavior of the prey also implies that differences in predator feeding mode may lead to species-specific differences in passive prey selection, even among closely related species. This is illustrated by pronounced differences in diet and feeding behavior of several species of co-occurring hydromedusae (Costello and Colin 2002, Colin et al. 2003). Hydromedusae with a more prolate-shaped bell, such as *Sarsia* considered above (figs 6.3B, 6.4), swim by jet propulsion. These species cannot feed while swimming, and prey encounters occur only via ambush of motile prey that swim into the outreached tentacles. Thus, these hydromedusae feed predominantly on motile prey, such as crustaceans and ciliated plankters, as also seen above for *Sarsia*. The more oblate hydromedusae swim by contracting the bell and thus "row" slowly through the water. These species feed and swim simultaneously, and nonmotile or slowly swimming prey are encountered as they are entrained in the feeding current produced by the "paddling" bell. These species feed mainly on soft-bodied prey, such as fish eggs and appendicularians, that have no or limited capability of detecting the hydrodynamic signal produced by the predatory medusa.

Prey selection based on chemical properties of the prey has been suggested, if not demonstrated, for nanoflagellates (Jürgens and De-Mott 1995), ciliates (Verity 1991), and copepods (DeMott 1988), among others. Landry et al. (1991), for example, showed that the interception-feeding 7-μm-diameter heterotrophic microflagellate *Paraphysomonas* clears living *E. coli* bacteria at much higher rates (by a factor of twenty) than heat-killed bacteria, and argued that the flagellate can discriminate between living and dead bacterial prey by chemical cues. However, chemically based selection between very small bacterial prey is not likely because diffusible chemical signals attenuate almost instantaneously (time scale milliseconds) on such small spatial scales (micrometers) as a result of molecular diffusion, as discussed in chapter 2. Theoretical considerations suggest that there is a particle size limit of about 2 μm below which chemical detection does not function (Jackson

1987). As demonstrated in chapters 3 and 4, interception-feeding flagellates may depend on prey motility for prey encounters even when the flagellate creates a feeding current. A Sherwood number calculated for *Paraphysomonas* feeding on bacteria in fact predicts a difference in encounter volume rates between living (motile) and dead bacteria of the observed order. Thus, in this case there is no need to invoke chemical cues as a selection mechanism. The observation by Gonzáles et al. (1993) that heat killing of bacterial prey leads to lower flagellate clearance rates only in motile bacteria supports this conclusion. The lack of a convincing example of chemically based prey selection in predators feeding on very small prey does not imply the lack of chemosensory capability. In fact, chemosensory behavior similar to that described for bacteria is widespread among protists, but this behavior must rather be considered an adaptation to aggregate at food patches than to selection between individual prey particles (Fenchel and Blackburn 1999).

Chemical selection may function for larger prey. It is well documented that heterotrophic dinoflagellates, for example, use chemical signals to locate phytoplankton prey (Jacobsen and Anderson 1986, Strom and Buskey 1993), and this may also be used as a means of selectively feeding on those prey species that, irrespective of size, provide the strongest or most "edible" chemical signal (Buskey 1997). Leaking phytoplankton cells are surrounded by a "phycosphere," as described in chapter 3, and predatory dinoflagellates respond when encountering the phycosphere by initiating prefeeding circling around the prey cell and eventually capturing it. Some copepods are able to selectively feed on fast-growing, more nutritious cells over slower-growing ones of the same species in mixtures (Cowles et al. 1988) or to select nontoxic over toxic cells of otherwise identical dinoflagellates (Teegarden 1999), in both cases presumably through chemically based prey selection. The latter may have implications for the dynamics of blooms of toxic algae. The mechanisms of chemically based prey selection in copepods are not clear, but some studies have suggested remote identification of prey cells (Koehl and Strickler 1981). When the cell enters the accelerating feeding current of a copepod, the phycosphere surrounding the cell will be drawn out, thus potentially allowing the copepod lead time to decide whether to grab a particular cell before it arrives (Andrews 1983, Moore et al. 1999). This suggested mechanism remains to be verified. In a similar but different fashion, zooplankters and bacteria may decide whether or not to track down and colonize a larger particle, depending on the chemical signature it leaves in its wake (see fig. 3.8).

6.5 PREY SWITCHING

Prey selectivity may not be a constant property but may vary with prey concentration and with the availability of alternative prey. Such prey selection may be considered "active," as feeding in mixtures cannot be predicted solely from single-prey experiments. One can distinguish between concentration-dependent and frequency-dependent active prey selection. Optimal foraging theory would predict that a predator should become more selective, i.e., pick the best, when prey is abundant and less selective when food is scarce. Such concentration-dependent selectivity has been demonstrated in a variety of planktonic predators, ranging from heterotrophic nanoflagellates (Jürgens and DeMott 1995) and copepods (DeMott 1989) to larval fish (Munk 1997). In most cases, the mechanisms are unresolved. Frequency-dependent selection, that is, prey selection dependent on the relative frequency of a particular prey, is also termed prey switching. Positive prey switching implies that the clearance rate of a particular prey increases with its relative abundance. As mentioned above, prey switching is another mechanism that may lead to an S-shaped type III functional response. Prey switching has been demonstrated in copepods (Landry 1981, Kiørboe et al. 1996, Gismervik and Anderson 1997) and a few other aquatic invertebrates (Lawton et al. 1974, Butler and Burns 1991, Elliott 2004, 2006) but may occur more commonly. The mechanism behind prey switching has been revealed in the copepod *A. tonsa* (Jonsson and Tiselius 1990, Kiørboe et al. 1996). As described in section 4.7, this copepod can feed in two different modes: ambush feeding and suspension feeding (fig. 4.11). In the suspension-feeding mode, prey entrained in the feeding current may attempt an escape, as described in chapter 5. Ciliates and many other protozoans are able to perceive the fluid deformation and to escape, whereas diatoms and other nonmotile phytoplankters are not. Conversely, in the ambush-feeding mode, motile prey are perceived by the hydrodynamic disturbance that they generate in the water, but nonmotile prey are neither perceived nor encountered. The copepod adopts the feeding mode that generates the best turnout, that is, ambush feeding when motile prey dominate, and suspension feeding when diatoms dominate. This leads to disproportionate feeding on the most abundant prey, that is, prey switching.

6.6 BIOENERGETICS: CONVERSION OF FOOD TO GROWTH
   AND REPRODUCTION

Ingested food needs to be digested and assimilated before it can be combusted or used for growth or reproduction. Digestion essentially implies

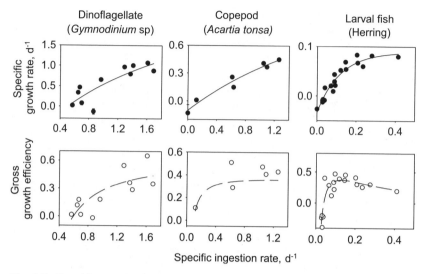

Fig. 6.7. Growth rates and gross growth efficiencies in various zooplankters as functions of ingestion rates: a dinoflagellate (A), a copepod (B), and a fish larva (C). Data were taken from Jakobsen and Hansen (1997), Berggreen et al. (1988), and Kiørboe (1989), respectively.

breaking down prey mechanically and enzymatically in the gut or—for unicellular organisms—in food vacuoles. Breakdown products are small molecules, typically simple sugars, small oligomers of amino acids, and fatty acids. These smaller molecules are then assimilated, that is, taken up via the gut epithelium or across the cell membrane. Parts of the food that cannot be broken down to constituents that can be assimilated are egested, often as feces. The assimilated molecules are either combusted to fuel synthesis and maintenance or again build into larger molecules, including biomass building blocks (fat, proteins, carbohydrates) and excretory products (e.g., urea). Synthesis of biomass implies growth, which in unicellular organisms typically is measured as an increase in cell concentration over time in experimental cultures; in metazoans it is typically quantified as a mass increment in individuals.

At all the steps in the transformation, from ingestion to final biomass synthesis, there are losses of matter and energy, implying that the growth efficiency or growth yield is considerably less than 100 percent. Empirically derived gross growth efficiencies in individuals (growth/ingestion) or growth yields in cultures (determined in batch cultures as mass of predators harvested divided by mass of prey initially added) for various zooplankters are typically in the range 30–50 percent (fig. 6.7). Gross growth efficiencies vary with prey availability and, thus, feeding rates and may be

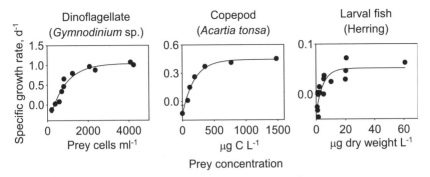

Fig. 6.8. The functional response in growth rate to prey concentration in a heterotrophic dinoflagellate (A), a copepod (B), and a larval fish (C). Data from Jakobsen and Hansen (1997), Berggreen et al. (1988), and Kiørboe and Munk (1986), respectively.

lower than typical rates at very low and at very high prey densities. The resulting growth rates typically vary with prey concentration in much the same way as the functional response in ingestion rate to prey concentration and in most cases can be fitted well by hyperbolic functions (fig. 6.8).

The efficiencies of some of the transformation steps are very much dependent on the composition of the food, and others can be constrained theoretically. The efficiency by which ingested food is assimilated varies; food containing a large fraction of material that cannot be digested (e.g., cellulose) is, of course, assimilated less efficiently than food consisting mainly of digestible material. Assimilation efficiency in zooplankters feeding on micro- or nanoplankton is typically about two-thirds but may be considerably less when they are feeding on "detritus." The fate of assimilated material is often more predictable: as much as possible goes toward growth, either in the form of soma or in the form of reproductive products. However, a certain fraction is allocated to "maintenance," and there are unavoidable biochemical costs of growth that put an upper bound on the efficiency by which assimilated food can be converted to growth (net growth efficiency). These latter costs can be measured experimentally as an increase in the metabolic rate following feeding ("specific dynamic action"), or they can be estimated theoretically.

6.7 SPECIFIC DYNAMIC ACTION: EGG PRODUCTION EFFICIENCY IN A COPEPOD

Consider as an example egg production in a female copepod (Kiørboe et al. 1985). Assimilated food has to be transported across cell membranes, and small molecules have to be combined into protein, fat, and complex carbohydrates to form an egg. Protein is formed from amino acids that are linked together by peptide bonds, and this, in particular, requires energy: it takes about four molecules of ATP to form one peptide bond. Because a copepod egg contains about 55 percent protein, and because the molecular weight of amino acids is on average 112, then ($0.55/112) \times 4 = 19.6 \times 10^{-3}$ µmol ATP is required to synthesize the protein contained in 1 µg (dry mass) of copepod egg. The rest of the egg is mainly lipids, which are much cheaper to synthesize from monomers, and one can calculate that it requires a total of about 23×10^{-3} µmol ATP to synthesize all the macromolecules in 1 µg of egg. In a similar fashion, one can estimate the "delivery costs," i.e., the biochemical costs of transporting monomers across cell membranes (about 7×10^{-3} µmol ATP to transport the constituents of 1 µg of egg). The resulting 30×10^{-3} µmol ATP for each microgram of egg formed has to be synthesized. This requires the combustion of some of the assimilated food, which in turn can be measured as oxygen consumption. By using appropriate conversion factors, about 6 nl O_2 is consumed for each nanomole of ATP synthesized. Thus, one would predict from these biochemical considerations that the metabolic rate of a copepod would increase by $6 \times 30 = 180$ nl O_2 for each 1 µg of egg produced. This is very close to that actually observed in *Acartia* [197 nl O_2 (µg egg produced)$^{-1}$]. This, in turn, suggests that *Acartia* produces eggs at an efficiency that is near its theoretical maximum and that the main cost of growth is protein synthesis. This has subsequently been demonstrated more directly (Thor 2000) and implies that the costs of feeding (e.g., generating feeding currents) are negligible in copepods, as has also been argued for protozoans (see Fenchel 1986). Similar conclusions have been arrived at for growth in planktonic fish larvae (Kiørboe et al. 1987, Kiørboe 1989).

The maximum possible efficiency by which assimilated energy in excess of maintenance requirements is converted into egg mass estimated from the above considerations is about 0.85 (when everything is converted to carbon units) and may in general vary between 0.8 and 0.9 depending on the composition of the food and the deposited biomass. Actual realized net growth efficiencies (growth/assimilation) are lower because some of the assimilated energy goes toward maintaining the organism. Realized net growth efficiencies in *Acartia,* for example, vary between 0.6 and 0.76, and the theoretical maximum (0.86) is approached only at the highest feeding rates, where maintenance becomes a smaller fraction of the total energy requirements of the organism.

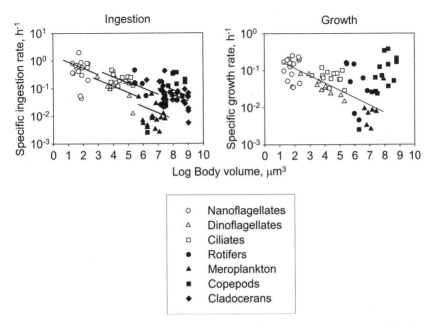

Fig. 6.9. Scaling of grazing and growth rates with body size in various planktonic predators. The scaling is similar among groups (slopes ~ −0.23), although the magnitudes of ingestion and growth rates vary somewhat. Modified from Hansen et al. (1997).

6.8 SCALING OF FEEDING AND GROWTH RATES

Specific growth rates that can be attained by small organisms such as bacteria (perhaps up to 5 d^{-1}) are very much higher than the maximum growth rates in larger organisms such as copepods (0.5 d^{-1}) and fish larvae (0.05 d^{-1}). The same trend applies to specific feeding rates: small organisms have much higher maximum feeding rates than large organisms. Thus, here again, we realize that size matters. It turns out that most vital rates, including feeding, metabolism, growth, and reproductive rates scale with body mass in much the same way, i.e., with body mass raised to a power of about 3⁄4. This implies that the specific rates (rate/body mass) scale approximately with body mass to a power of −1⁄4 (fig. 6.9). This kind of scaling applies not only to planktonic organisms but much more generally, although the lead coefficients may vary among groups of organisms, as also seen for grazing rates among planktonic predators (fig. 6.9). This size scaling of vital rates implies that the significance of the organisms in an ecosystem in terms of production and turnover rates cannot be judged simply from their biomass: small organisms are relatively more important than their biomass contribution would suggest, and large organisms are similarly less important.

6.9 Feast and Famine in the Plankton

Plankton live in a heterogeneous world, and the availability of food varies on all spatial and temporal scales. We have previously seen how millimeter-scale solute plumes may develop around suspended or sinking particles that leak (figs. 3.3 and 3.6). Similarly, ephemeral solute plumes of a few minutes' duration may develop where a cell lyses or a copepod excretes (Jackson 1980, Blackburn et al. 1998). This is the kind of heterogeneity that bacteria and microplankters may experience. At somewhat larger scales, horizontal layers with elevated concentrations of phytoplankton and other organisms, extending a few centimeters to a few meters in the vertical direction, are found in vertically stratified regions, typically near the pycnocline (Dekshenieks et al. 2001, Cowles et al. 1998). Localized regions of elevated phytoplankton concentrations at the kilometer scale in the horizontal may occur around oceanic fronts, and ephemeral phytoplankton blooms lasting for some days or weeks occur at the onset of water-column stratification during spring in many temperate oceans and subsequent to upwelling or wind-mixing events (chapter 8). These "patches" of food—in time or space—may contrast with vast areas or periods with low food availability. The plankton has to cope with all this heterogeneity. There are numerous examples of physiological, biochemical, and behavioral adaptations to such a "feast and famine" existence (Koch 1971) that allow plankters to survive periods with little or no food and to rapidly take advantage of food that becomes available locally and/or ephemerally.

Optimal survival strategies in heterogeneous environments depend on the length of expected starvation periods or on the scale of the patchiness relative to the motility of the organism. We saw above how various zooplankters are capable of feeding and growing at a range of food concentrations and how their metabolic expenses are reduced when prey availability is low; this may be considered an adaptation to moderate and short-term variability in food availability. Similarly, variation in motility patterns with food availability may improve the organisms' chances of finding food and of remaining in food patches. For example, many plankters adopt a more directionally persistent swimming pattern in the absence of food, which enhances the encounter rate with food patches (fig. 6.10A), and they reduce their motility in the presence of food, which helps them remain in patches (fig. 6.10B) (see also section 4.10). Examples

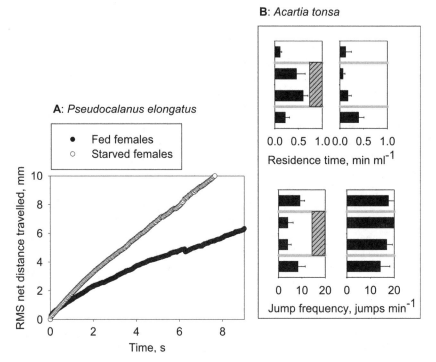

Fig. 6.10. Motility and food availability in zooplankton: two examples of the effects of food availability on motility. A: Females of the copepod *Pseudocalanus elongatus* in the absence and presence of food: *RMS* net distance traveled as a function of time (see section 2.2). Swimming velocity of the females is invariant with food availability (~2 mm s^{-1}), but in the absence of food, the motility of the females become more directionally persistent, and, as a result, they have longer equivalent "run lengths" and higher diffusivities (by a factor of six), with the result that the chance of finding a patch of food is elevated. Original data. B: Residence time and motility (quantified as jump frequency) of a copepod (*Acartia tonsa*) in a three-layered water column (10 cm high) with or without phytoplankton in the middle layer. The layers are indicated by vertical lines, and the presence of phytoplankton by gray shading. The copepod increases its residence time in the layers with food by reducing its jump frequency (motility). Modified from Tiselius (1992).

of such adaptations in motility patterns are found among all kinds of planktonic organisms from protozoans (Bartumeus et al. 2003, Menden-Deuer and Grünbaum 2006) to copepods (Tiselius 1992) and larval fish (Hunter and Thomas 1974). When entering a patch of food, some copepods may compensate for a previous starvation period by elevated feeding rates immediately subsequent to food exposure (Tiselius

1998) and return to "steady-state" feeding rates after some time, much the same way as we saw above for the little *Sarsi* jellyfish with the voluminous gut (fig. 6.2A,B).

Some pelagic bacteria appear to specialize in attaching to surfaces such as those provided by suspended particles (e.g., Riemann et al. 2000), but the concentration of large, attractive marine snow particles may be low, and the distance between them consequently substantial compared to the swimming velocity of a bacterium. At typical oceanic size distributions and concentrations of particles and with typical motilities of bacteria, characteristic encounter rates are on the order of one particle per bacterium per day (Kiørboe et al. 2002). That is, the search time is on the order of 1 d. Are pelagic bacteria capable of swimming that long without food? The cost of swimming in bacteria is low and can be estimated from the viscous drag multiplied by the swimming velocity as $6\pi\varsigma a v^2$, where ς is the dynamic viscosity ($\sim 10^{-2}$ g cm^{-1}s^{-1}) (Berg 1993). A 1-μm^3 volume (0.6-μm radius) bacterium swimming at 100 μm s^{-1} thus requires power at 10^{-16} J s^{-1} (recall that 1 J = 1 Nm = 1 kg m^{2}s^{-2} = 10^7 g m^{-2}s^{-2}). By metabolizing its own cell mass, the bacterium may release about 0.5×10^{-8} J, thus allowing it to swim continuously for 0.5×10^{-8} J/10^{-16}J s^{-1} = 5×10^7 s $\approx$ 600 d before it burns out. Even if the conversion of chemical to mechanical energy is very low, e.g., 1 percent, a searching bacterium would on average have sufficient time to find a particle before it runs out of energy.

When food availability becomes even lower and starvation periods longer, many planktonic organisms may turn down their cellular machinery and metabolism, much more than the above considerations of "cost of growth" would imply. This is particularly pronounced in bacteria (Koch 1971) and protozoans (Fenchel 1986), where starvation respiration may become only a few percent of the respiration of feeding and growing cells; but even in copepods, fasting respiration may decline to less than 25 percent of that of actively feeding individuals (Kiørboe et al. 1985). Many microplankters, including heterotrophic dinoflagellates (Menden-Deuer et al. 2005) and other microflagellates (Fenchel 1986), can consequently survive for months without food and are capable of almost immediately initiating feeding and biomass deposition when food again becomes available. Larger zooplankters, such as copepods, may build up lipid reserves to allow survival during winter (Hirsche 1996). A more dramatic change in the physiology of an organism is encystment, which is found both among phytoplankton and heterotrophic protozoans. In fact, many microorganisms that are considered "pelagic" spend most of their lives in the sediment. Pelagic diatoms, for example, appear well adapted to long-lasting, slow life in the sediment. Subsequent to short-lived blooms, most of the cells sink to the seafloor, often in the form of marine snow aggregates (chapter 4), and concentrations of viable diatom spores in marine sediments may exceed millions of cells per square centimeter (Hansen and Josefson 2003). These spores survive consumption by deposit-feeding invertebrates (Hansen and Josefson 2004) and may seed the water column when conditions again become favorable for their growth. Resting stages may occur even among larger zooplankters, for example copepods, that can produce resting eggs. These eggs may survive in the sediments for decades (Katajisto 1996) or even for centuries (Hairston et al. 1995) and provide an abundant (up to $>10^6$ eggs m^{-2}) source for recruitment to the water column (Katajisto et al. 1998).

Chapter Seven

POPULATION DYNAMICS
AND INTERACTION

7.1 FROM INDIVIDUAL TO POPULATION

A *POPULATION* IS A group of individuals belonging to the same species that can somehow be separated from other groups of the same species, typically by having a different and distinct spatial or temporal distribution. Populations of planktonic organisms are, for obvious reasons, often difficult to define exactly. By *population dynamics* we mean how the population composition and population size vary in time (and space). Essentially, the dynamics of a population depends on two parameters, the birth rate and the death rate (mortality), with the difference between the two being the population growth rate. Birth and death rates are features of populations, not of individuals. One cannot, for example, talk about the death rate of an individual—it either dies or does not during a particular time interval. Still, though, population parameters are related to events at the individual level—they are statistical summations of individual-level events. The dynamics of the population is thus the result of interactions occurring at the level of the individual. In particular, population processes depend on encounters between individuals, whether they are encounters between mates, between predators and their food, or between prey and their predators. These encounters determine the fate of the individual and sum up to govern birth and death rates of the population.

In the following chapter we examine some properties of the dynamics of populations of plankton, and, whenever possible, try to utilize the insights achieved in previous chapters about individual encounter rates to make predictions about populations. We first consider single populations, where we do not worry too much about what causes birth and death rates to vary. Subsequently, we consider the covariation between populations, e.g., between predator and prey populations or between competing populations. Throughout we apply well-established population theory and (variants of) classical methods and models such as life-table analysis and Lotka-Volterra-type models to populations of plankton in an attempt to make predictions about pelagic populations and to explain observed patterns.

7.2 The Dynamics of a Single Population: Phytoplankton Blooms

Populations may grow at a constant specific rate whenever conditions are favorable and unlimiting. A constant, positive growth rate implies an exponentially increasing population size

$$\frac{dC}{dt} = \mu C \Rightarrow C_t = C_0 e^{\mu t} \tag{7.1}$$

where C_0 and C_t are the organism concentrations at times 0 and t, and μ is the specific growth rate. Examples of exponential population increase include laboratory cultures of single cells (fig. 7.1A,B); here μ is simply the number of cell divisions per cell per unit time, equivalent to the "birth rate." From the ocean, the initial phase of a phytoplankton bloom offers another example of exponential population increase.

Obviously, however, exponential growth cannot continue forever; eventually some resources may become limiting, growth ceases, and the population density may approach a more or less constant level (fig. 7.2). The saturating response in population development is often modeled with the logistic equation. Logistic growth assumes that the population growth rate declines linearly with increasing population density and is zero when the population reaches the carrying capacity of the environment. Specifically, it is assumed that the growth rate is proportional to

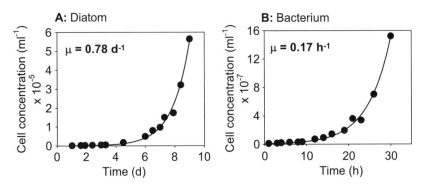

Fig. 7.1. Growth of a marine pelagic diatom and a marine pelagic bacterium in laboratory cultures.

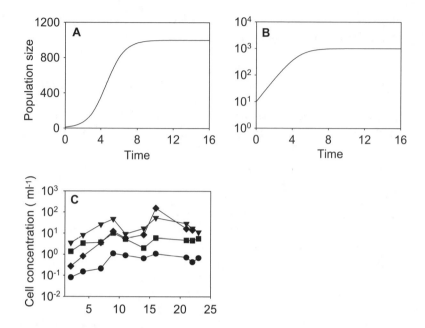

Fig. 7.2. Schematics of logistic growth on linear (A) and log scales (B) and example of diatom population development (five species, each with its own symbol) in a shallow Danish fjord during spring (C). Panel C is based on data in Kiørboe et al. (1994).

the relative difference between the carrying capacity (M) and the current population density, i.e.,

$$\mu(C) = \mu\left(\frac{M-C}{M}\right) \tag{7.2}$$

Substituting $\mu(C)$ for μ in equation 7.1 yields

$$\frac{dC}{dt} = \mu\left(\frac{M-C}{M}\right)C = \mu C - \frac{\mu}{M}C^2$$

$$\Rightarrow C_t = \frac{M}{1 + be^{-\mu t}} \tag{7.3}$$

where b is a dimensionless constant. Note that the logistic equation is only descriptive and that there are no real mechanisms embedded in it the way we have formulated it here. For phytoplankton, for example, the limiting resource can be carbon dioxide, nutrients, or light (self-shading), or the cells may be restrained by the increasing pH that is a result of the consumption of carbon dioxide during photosynthesis (Hansen 2002). For field populations (fig 7.2C), there are several additional possible factors that may slow down population increase, such as grazing or aggregation and subsequent sedimentation. Thus, it is generally difficult to tell much about the factors that regulate the growth rate of an organism alone from observations of the growth curve. In the following we consider two examples of logistic-like population growth in phytoplankton in a bit more detail, namely the effects of coagulation and self-shading on population development.

7.3 Phytoplankton Population Dynamics and Aggregate Formation

The general and descriptive logistic growth equation can be given a mechanistic interpretation in specific cases. One example is phytoplankton coagulation. *Coagulation* is the process by which solitary phytoplankton cells stick together on collision to form aggregates that subsequently may sink out of the water column, as discussed in section 4.4. Because the encounter rate between suspended cells in a monospecific bloom increases with the square of the cell concentration (see eq. 4.4), aggregation and sedimentation rate accelerate during a phytoplankton bloom and hence slow down population increase. Jackson (1990) and Jackson and Lochmann (1992) combined phytoplankton growth kinetics and coagulation theory in a simple population dynamic model. In its simplest form, only collisions between single cells are considered; even when large aggregates are formed, this is the quantitatively most important process. If we let the phytoplankton cells both coagulate and grow, then

$$\frac{dC_1}{dt} = \mu C_1 - \alpha\beta C_1^2 \tag{7.4}$$

where the first term on the right-hand side describes the phytoplankton growth (μ is the specific growth rate), and the second term is the cell loss to coagulation and subsequent sedimentation (see eq. 4.5; recall that α is the stickiness of the algae and β the encounter rate kernel for coagulation). Note that equation 7.4 is exactly the logistic equation (eq. 7.3), but the model now has a mechanistic underpinning. The population dynamics of five species of diatoms during a spring bloom shown in Fig. 7.2C can in fact be explained by coagulation and equation 7.4. This does not follow solely from the observed dynamics but is supported by simultaneous observations of coagulation rates and sedimentation rates (see fig. 4.8) as well as other evidence (see below). The model predicts sigmoid

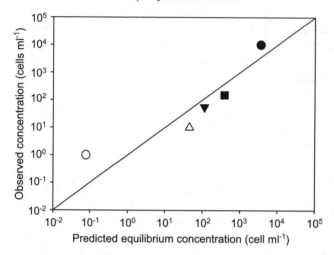

● *Skeletonema costatum*
○ *Coscinodicus concinnus*
▼ *Thalassiosira* sp.
△ *Rhizosolenia h.f. semispina*
■ *Leptocylindricus danicus*

Fig. 7.3. Observed and predicted (eq. 7.5) equilibrium concentrations of diatoms in a Danish fjord during a spring phytoplankton bloom. The line is $X = Y$. The three parameters used for predicting equilibrium concentrations were estimated from experimentally determined stickiness coefficients (α), from wind-based estimates of turbulent shear combined with observed cell sizes (β), and from observations of growth rates at low cell concentrations (μ). Modified from Kiørboe et al. (1994).

population growth, even when light and nutrients are nonlimiting (constant growth rate). As the population grows, coagulation becomes increasingly important, and there will be a steady-state critical cell concentration (C_{Cr}) where growth and coagulation balance one another and the population stops increasing. This concentration can be found by putting $dC_1/dt = 0$ and solving equation 7.4 for C:

$$C_{Cr} = \frac{\mu}{\alpha\beta} \tag{7.5}$$

This simple model predicts fairly accurately the diatom equilibrium concentrations found during a spring phytoplankton bloom in a Danish fjord (fig. 7.3). Thus, coagulation may control phytoplankton population dynamics.

7.4 PHYTOPLANKTON GROWTH AND LIGHT LIMITATION

Another example of sigmoid population growth where the underlying mechanisms can be described is light-limited phytoplankton growth and self-shading (modified from Huisman et al. 1999a and Huisman 1999). This example also forms the basis for later examining interspecific competition for light. Consider a well-mixed water column of depth h with a nonlimiting concentration of inorganic nutrients. As a result of the presence of phytoplankton, light attenuates in the water column, and the phytoplankton growth rate will vary with light intensity and, hence, depth (z). The rate of change in population concentration, C, averaged over the water column is

$$\frac{dC}{dt} = \frac{1}{h} \int_{z=0}^{h} \mu[I(z)] dz - \sigma C \tag{7.6}$$

where σ is the depth-independent specific mortality rate, and $\mu[I(z)]$ is the specific growth rate as dependent on the light intensity, I, that varies with depth, z. Equation 7.6 simply states that the rate of change in population concentration is the difference between the growth rate averaged over the water column (first term on right side) minus the mortality rate (second term). The light intensity declines exponentially with depth as follows (Lambert-Beers law):

$$I(z) = I_0 e^{-kCz} \tag{7.7}$$

where k is the species-specific light attenuation coefficient (dimensions of L^2). The phytoplankton specific growth rate is assumed to vary with light intensity according to a simple Monod expression (note the similarity to the disk equation, eq. 6.4):

$$\mu(I) = \frac{bI}{1 + b\mu_{max}^{-1}I} \tag{7.8}$$

where μ_{max} is the maximum specific growth rate, and b is the initial slope of the μ versus I curve. Under the specified conditions, the population will continue to increase until self-shading brings the water-column averaged growth rate in balance with the mortality rate, and the population will reach a steady-state concentration, C^*, as growth ceases. One can define the "critical light intensity" as the light intensity at the bottom at steady state, $I_{Cr} = I_0 \exp(-kC^*h)$. This set of coupled equations together predicts a sigmoid growth curve similar in shape to logistic growth (fig. 7.4A,B), and its components (assumptions) can be examined and quantified separately (e.g., μ–I curves and light attenuation) in experimental

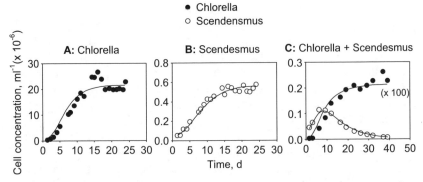

Fig. 7.4. Population development of two phytoplankton species in monoculture (A, B) and mixed-culture (C) experiments. The solid lines in panels A and B are fits of the self-shading model (eqs. 7.6–7.8) to the data: self-shading leads to a sigmoid population development and a steady-state population density when self-shading prevents net growth. In panel C parameters estimated from the monoculture experiments were used to predict population development in mixed-species experiments (see section 7.9.3). Modified from Huisman et al. (1999a).

systems. This in principle allows one to predict population development from estimates of b, k, σ, and μ_{max}.

7.5 Scaling of Growth and Mortality Rates

Potential population growth rates differ among species and generally are larger in small than in large organisms. Potential population growth scales with organism size much the same way as the growth rates of individuals and other vital rates (feeding rate, respiration rate, etc.) of individuals scale with size, i.e., approximately with body mass raised to a power between $-1/3$ and $-1/4$ (fig. 7.5A). Potential growth rates, however, are rarely realized or are realized for only short periods, either because resources become limiting at high population densities (see above) and/ or because mortality kicks in. The realized population growth rate, μ, is the difference between birth or cell division rate and mortality rate. Because, on average, populations have growth rates of 0—otherwise they would grow to infinite size or go extinct—mortality rates must be of similar order as population growth rates and scale with organism size in the same way. This, in fact, is the case, as demonstrated in figure 7.5B for planktonic organisms.

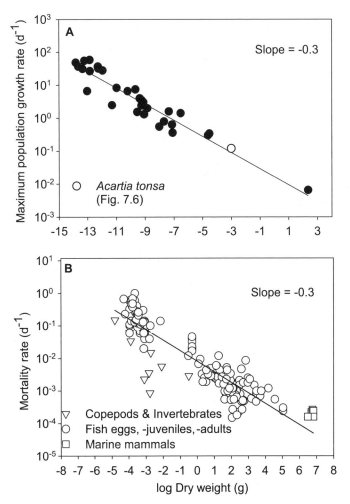

Fig. 7.5. Maximum specific population growth rates of bacteria, protists, and pelagic invertebrates as a function of organism size (A) and mortality rates of pelagic organisms as a function of their body mass (B). Note that mortality and maximum population growth rates scale similarly with organism size and are of similar magnitudes. Data in A from Fenchel (1974), and those in B from McGurk (1986). The maximum population growth rate of the copepod *Acartia tonsa* estimated in table 7.1 and figure 7.6 is included in panel A (open symbol).

7.6 POPULATIONS WITH AGE STRUCTURE: LIFE TABLES

The considerations above apply only to organisms with continuously over-lapping generations and in cases where all individuals can be considered identical, as would be approximated by unicellular organisms. For species with more complex life histories and more or less distinct generations, one can still define a population growth rate, but it cannot be computed simply from counting individuals. Here we appeal to "life tables" to compute population growth rates. Life tables can also be used to examine various life-history characteristics and adaptations. We briefly introduce life tables and then apply the approach to examine the life histories of pelagic copepods, the absolutely dominating zooplankton group in the ocean.

A life table is a simple way of actuarially bookkeeping the individuals in a cohort (table 7.1) that can be illustrated graphically (fig. 7.6). From a life table one can estimate the average number of offspring that an average individual delivers to the next generation as well as the population growth rate. We have copepods in mind, but the approach applies generally. Assume that we start with a cohort of eggs with age $x=0$. As the eggs hatch and develop into nauplii and, subsequently, copepodites, their numbers will decline because of mortality. We describe this by their age-dependent survival, l_x (fig. 7.6A). We can similarly define the age-dependent fecundity, m_x (female egg production) (fig. 7.6B). We follow the cohort until all its members have died. The number of offspring that an average female egg delivers to the next generation, the net reproductive rate R_0, is then

$$R_0 = \sum_x l_x m_x \tag{7.9}$$

which is illustrated by the area under the curve in figure 7.5C. If, instead, we consider age classes of infinitesimal duration, then

$$R_0 = \int_0^\infty l_x m_x dx \tag{7.10}$$

Population maintenance requires that each female on average delivers one male and one female to the next generation, i.e., that R_0 is at least two. If it is less, the population will decline and eventually go extinct. Conversely, if R_0 exceeds two, then the population will increase. Averaged over long times, R_0 for any (closed) population must be two—if it were less, the population would not exist, and if it were more, it would increase infinitely.

Assuming a 1:1 sex ratio, then $R_0/2$ is the factor by which the population will increase (or decrease) in the course of one generation. If the generation time is T, then the population growth rate is

TABLE 7.1
Life Table for a Laboratory Population of Copepods, *Acartia tonsa*, at 18°C

Stage	Age (weeks) x	Survival l_x	Fecundity (eggs female^{-1} week^{-1}) m_x	$l_x m_x$	$x l_x m_x$
Egg	0	1	0	0	0
NI-III	0.5	0.61	0	0	0
NIV-CI	1.5	0.43	0	0	0
CII-CV	2.5	0.30	0	0	0
CVI	3.5	0.25	140	35	121
CVI	4.5	0.20	287	58	261
CVI	5.5	0.17	189	31	172
CVI	6.5	0.14	147	20	129
CVI	7.5	0.11	133	15	111
CVI	8.5	0.09	112	10	86
CVI	9.5	0.07	80	6	56
CVI	10.5	0.06	0	0	0
			$\Sigma m_x = 1088$	$R_0 = \Sigma l_x m_x = 174$	$\Sigma x l_x m_x = 936$

Constructed from data in Parrish and Wilson (1978), Kiørboe et al. (1985), and Berggreen et al. (1988).

$$\mu = \frac{\ln(R_0/2)}{T} \qquad (7.11)$$

The generation time can be estimated as the average age of the mother of a random egg

$$T = \frac{\int_0^\infty x l_x m_x dx}{\int_0^\infty l_x m_x dx} \qquad (7.12)$$

In table 7.1 and figure 7.6 we consider as an example the life table of a laboratory culture of the copepod *Acartia tonsa*. The integrals in equation 7.10 can be approximated by the summations (Σ). The lifetime egg production of a female (Σm_x) is more than 1000, and the net reproductive rate ($R_0 = \Sigma l_x m_x$) of 174 implies that the population will increase by a factor of $174/2 = 87$ in one generation. Because the generation time can

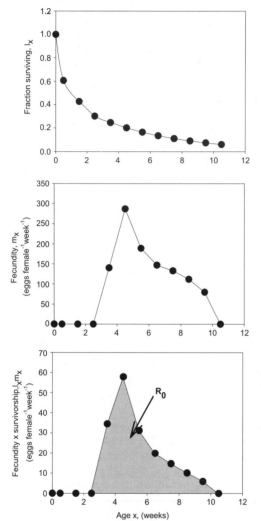

Fig. 7.6. Graphic illustration of life table data for the copepod *Acartia tonsa* (see table 7.1). The upper panel shows the survivorship curve, and the middle panel the age-dependent egg production rate. The lower panel plots the product of age-dependent survivorship and fecundity. The shaded area under this curve is an estimate of the net reproductive rate.

be estimated as $T = \Sigma x l_x m_x / \Sigma l_x m_x = 936/174 = 5.4$ weeks, this implies a population growth rate of $\mu = \ln(174/2)/5.4$ week$^{-1} = 0.83$ week$^{-1} = 0.12$ d^{-1}. This growth rate is much higher than one can expect to find in field populations, even during short time intervals, because mortality rates there are much higher—because of predation—than in a laboratory population. For example, the highest population growth measured in coastal populations of small copepods was 0.025 d^{-1} (Kiørboe and Nielsen 1994).

7.7 BEHAVIOR AND POPULATION DYNAMICS: CRITICAL POPULATION SIZE AND ALLEE EFFECTS

In the above we have described various properties of a population and introduced tools to make these descriptions, but we have not considered the mechanisms that cause mortality and fecundity and, hence, population size to vary. Such mechanisms operate at the level of the individual. In the following we again use copepods to illustrate how individual behaviors govern population processes.

We have hitherto assumed that the fecundity of a female copepod is determined only by its age, but it may also be constrained by the availability of males—the female needs to be mated in order to produce (fertile) eggs. This, of course, applies generally to all organisms with sexual reproduction and in particular to planktonic organisms because they live in a three-dimensional world where finding mates may represent a particular challenge. This, in turn, implies that there must be a lower critical population density below which the mate-encounter rate simply becomes insufficient to allow maintenance of the populations (Gerritsen 1980). It also implies that even above this critical population density, the population growth rate will depend on the population density—the higher the density, the higher the rate of mate encounters and therefore the higher the fecundity and population growth rate (at least until a density where mating frequency becomes saturated). That is, population growth rate becomes positively density dependent. This is termed the Allee effect (Stephens et al. 1999; note that there are other processes that can lead to Allee effects). We can use the pelagic copepod example to estimate the magnitude of both the critical population density and the Allee effect.

Assume now that the fecundity of a female depends on the probability that it has been mated at age x (p_x) and therefore that the fecundity is the product of this probability and the rate of egg production (f), which we assume to be age independent

$$m_x = p_x f \tag{7.13}$$

The probability that a female has been mated at age x depends on both the concentration of males (C_M) and the capacity of males to find receptive females. The latter can be expressed as a volumetric encounter rate, β, as discussed in chapter 3. Females will encounter males at a rate βC_M. Assuming that mate encounters occur randomly in time (a Poisson process), the probability that a female has been mated at least once at age x is then

$$p_x = (1 - e^{-\beta C_M (x - \kappa)}), \text{ for } x > \kappa \tag{7.14}$$

where κ is the age of maturation, and $(x-\kappa)$ is thus the duration of time that a female of age x has been adult and available for mating. Combining equations 7.10, 7.13, and 7.14 yields (details of the derivation are in Kiørboe 2006)

$$R_0 = l_x \frac{f}{\sigma} \left(\frac{\beta C_M}{(\sigma + \beta C_M)} \right)$$

(7.15)

where σ is the mortality rate subsequent to hatching of the egg (assumed to be constant). Recall that the requirement for population maintenance is that $R_0 \geq 2$. We can rearrange equation 7.15 to estimate the minimum concentration of males (C_M^*) required to allow population maintenance $(R_0 = 2)$

$$C_M^* = \frac{\sigma^2}{0.5\beta(l_x f - \sigma)}$$

(7.16)

Equation 7.16 says, in accordance with intuition, that the larger the capacity of the males to find females, quantified as β, the lower is the critical density that allows maintenance of the population. Similarly, high egg production capacity (f) of the females allows the population to be maintained at low densities.

Using typical vales of the parameters (mortalities, development times, and size-dependent mate encounter kernels, see fig. 3.10B), equation 7.16 provides a fair description of minimum adult population sizes of pelagic copepods in the North Sea (fig. 7.7). Thus, insights into individual mating behaviors, as discussed in chapter 3, allow us to predict densities of copepods in the ocean.

We can similarly examine how the population growth rate (eq. 7.11) depends on the adult density (fig. 7.8) using the expressions for generation time (eq. 7.12) and net reproductive rate (eq. 7.15). Again using typical parameter values for a small pelagic copepod, figure 7.8 shows how population growth rate is expected to vary with population density. This suggests that Allee effects may be important at low but common winter densities of pelagic copepods, that is, producing positively density-dependent population growth. This implies that the lower the population density, the lower the growth rate, and vice versa. Although Allee effects have never been directly demonstrated for any marine population (Gascoigne and Lipicus 2004), the present analysis may explain why year-to-year variations in winter densities of copepods in the North Sea become amplified the subsequent summer (Colebrook 1985). Low winter densities lead to low population growth rates and, hence, low summer densities.

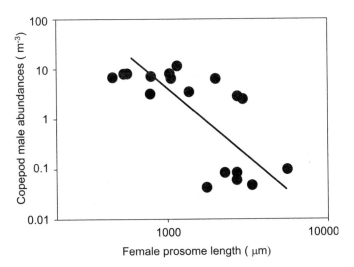

Fig. 7.7. Observed (dots) and predicted (line) minimum population sizes of adult males of pelagic copepods in the Skagerrak/North Sea. During the annual cycle in temperate waters, population sizes of pelagic copepods have their seasonal low in late winter/early spring. Redrawn from Kiørboe (2006).

The concepts of critical density and Allee effects of course apply much more generally, although it may not always be possible to explicitly quantify them. One implication of the critical density concept is that it constrains the diversity of sexually reproducing pelagic organisms because population densities cannot become infinitely small. If the density declines to below the critical density, the population will go extinct. This implies that the diversity of zooplankton must decline at low zooplankton biomass because there is room for only a limited number of species as each has to exceed its critical population density. This is consistent with the observation that zooplankton diversity in the ocean declines at low zooplankton biomass (Irigoien et al. 2004). In principle, it would be possible to provide crude estimates of this diversity, but in practice such estimates may be invalidated by connectivity and migration between populations. In fact, many planktonic organisms have circumpolar distributions.

7.8 LIFE-HISTORY STRATEGIES

We can use demographic analysis to explore the dynamics of planktonic populations as well as to examine life-history properties and optimal life-history strategies of planktonic organisms. This is because

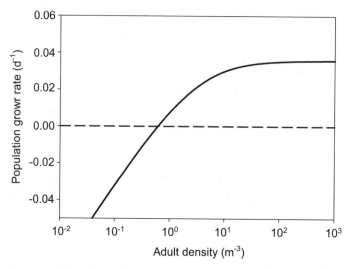

Fig. 7.8. Allee effect. Population growth rate versus adult density for a generic 1-mm pelagic copepod predicted from equation 7.11 (see text). Modified from Kiørboe (2006).

the net reproductive rate, R_0, in many cases can be interpreted as a measure or proxy of individual fitness (e.g., Stearns 1992)—R_0 estimates the average number of offspring that an individual will deliver to the next generation when we interpret l_x as a survival probability. Natural selection will favor those individuals that deliver the most offspring to the next generation. That is, individual R_0 is maximized by natural selection. We utilize this deduction below to examine reproductive and development strategies, again with pelagic copepods as example.

7.8.1 Reproductive Strategies in Copepods

Assume initially that the mortality rate, σ, is age independent; then the survivorship $l_x = e^{-\sigma x}$. Whether the assumption of age-independent mortality is realistic depends on the organisms under consideration. Many pelagic copepods spawn their eggs freely in the water (broadcast spawners), and these eggs suffer from a very high mortality rate; estimates of egg mortalities on the order of 1 d^{-1} or higher are not uncommon (Peterson and Kimmerer 1994). One reason for the high egg mortality rates is that copepod eggs are of a size suitable for many pelagic predators, including the copepods themselves. In contrast, the post-hatching stages (nauplii and copepodites) suffer much lower mortalities, on the order of

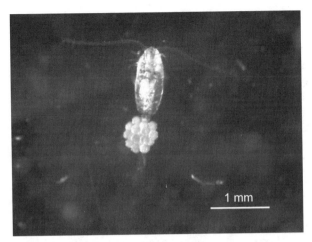

Fig. 7.9. *Pseudocalanus elongatus* carries its eggs, as do several other marine pelagic copepods. Other species spawn their eggs freely in the water, where they hatch. The two types are referred to as sac spawners and broadcast spawners.

0.1 d^{-1} (Hirst and Kiørboe 2002). For such copepods, we may therefore distinguish between two mortality rates, one that applies to the eggs and one that applies to all subsequent stages. In other species, the female carries the eggs in an egg sac until hatching (fig. 7.9), and the mortality of the eggs must be (almost) identical to the mortality of the females. Thus, in this case, it is reasonable to assume an age-independent mortality rate.

In the case of different mortality rates of eggs and post-hatch stages

$$l_x = e^{(-\sigma_e-\sigma)\zeta}e^{-\sigma x} \tag{7.17}$$

where σ_e is the mortality rate of the eggs, σ the mortality of all subsequent stages and ζ is the hatching time of the eggs. If we for simplicity assume that the egg production rate is age-independent once females have become mature, or, $m_x = m$ for $x > \kappa$, where κ is the time to maturation, then it can be shown that equations 7.10 and 7.17 combine and reduce to

$$R_0 = (m/\sigma)e^{-(\sigma_e-\sigma)\zeta-\sigma\kappa} \tag{7.18}$$

We can make several inferences from equation 7.18. First, for species where the egg mortality is higher than the post-hatch mortality ($\sigma_e > \sigma$), as would apply to broadcast-spawning copepods, R_0 declines nearly exponentially with egg-hatching time. Thus, one would expect strong selective pressure to reduce egg-hatching time in broadcast relative to sac spawners. In other words, because mortality is so high in the egg stage, the time as egg should be reduced in order to increase the fitness of the individual. This, in fact, appears to be the case, as egg-hatching times in broadcast spawners are only about one-third of that in sac spawners (Kiørboe and Sabatini 1994). Recently hatched copepod nauplii may use hydrodynamic signals (chapter 5) to perceive approaching predators, and the nauplii have quite impressive escape capabilities (Mauchline 1998). Thus, hatching increases the probability of escaping predators and reduces the mortality rate. Early hatching occurs at a price, though, because the early nauplii of broadcast spawners do not have the capability to feed (Landry 1983), in contrast to the early nauplii of sac spawners, which appear to hatch at a more developed state and are able to commence feeding right away (Uchima and Hirano 1986).

Again, these considerations apply more generally. For example, hatching as a nonfeeding stage is common among planktonic organisms. Thus, most fish with pelagic eggs hatch as nonfeeding yolk-sac larvae. And again, the larvae may escape predators where the eggs cannot. Predation rates on pelagic as well as benthic fish eggs may be intense (e.g., Köster and Mölmann 2000), and there is therefore a strong selective pressure for early hatching of larvae before they are fully functional.

Along the same line of reasoning, because freely spawned eggs have such high mortality risks, there is a particularly strong selection pressure to maximize the rate of egg production in broadcast spawners. As a result, broadcast spawners have egg production rates that are about three times those of sac spawners (Kiørboe and Sabatini 1995).

Females that carry their eggs are more susceptible to visual predators and may thus suffer from a higher mortality than nonovigorous females (Sandstrom 1980, Bollens and Frost 1991). However, planktonic copepods appear to have evolved alternative ways of reducing their mortality. Thus, it has been demonstrated for several species that the ovigorous females reside deeper in the water column than non-ovigorous females of the same species (Vuorinen 1987, Bollens and Frost 1991). Here visual predation is limited as a result of reduced light, but food availability is also typically lower. Other species of egg-carrying copepods, such as *Oithona* spp., move very little and do not generate a feeding current (Paffenhöfer 1993), thus reducing their susceptibility to both visual and tactile predators. Again, this may reduce the rate of food acquisition and, consequently, growth and fecundity rates. Thus, because there is a risk associated with feeding, there is a trade-off between feeding on the one hand and survival on the other. In sac-spawning copepods, natural selection has lead to the evolution of a safer feeding strategy in order to enhance egg survival but at the cost of lower feeding and fecundity rates, whereas in broadcast spawners, feeding and fecundity rates are higher, which, in turn, is necessary to compensate for the elevated mortality rate of the eggs.

More generally, trade-offs between feeding and reproduction on the one hand and predator avoidance on the other may lead to distinct species- and age-specific vertical distributions and diel vertical migration patterns in zooplankton (Ohman et al. 1983, Frost and Bollens 1992, Titelman and Fiksen 2004). Such patterns may be predicted by optimizing the trade-off functions in simple life-history models in ways that are similar to the approach presented above (see Fiksen 1997).

7.8.2 Optimum Maturation Age

Evolution is an optimization game, as we saw above, and the question simply is which strategy allows an individual to deliver the highest number of individuals to the next generation. We can also use this question to examine the optimum age of maturation, again using copepods as the example. The general idea, though, also applies to other organisms (Kozlowski 1992). Most pelagic copepods grow through twelve stages (egg, six naupliar stages, and five copepodid stages) before eventually reaching adulthood, whereupon they do not molt and grow any further. Larger copepods produce more eggs than small ones, and even within a species, there may be substantial variation in sizes of adult females and in their fecundities (e.g., Ban 1994, Runge and Plourde 1996). Growth and developmental rates are not tightly coupled in copepods; individuals developing faster end up as smaller adults than individuals that develop slowly if they grow at the same rate. Rapidly developing individuals will have a higher chance of reaching adulthood than slowly developing ones because mortality operates during a shorter time period, but the rapidly developing individuals will end as smaller adults with lower egg production rates. Which is the optimum strategy in terms of average number of individuals delivered to the next generation (equivalent to R_0), the fast or the slow developer?

Assume that fecundity varies with the development time as a power function, i.e.,

$$m(\kappa) = b\kappa^c \tag{7.19}$$

where b is a proportionality constant and c the power. Equation (7.18) then modifies to:

$$R_0 = \frac{b\kappa^c}{\sigma} e^{-(\sigma_e - \sigma)\varsigma - \sigma\kappa} \tag{7.20}$$

The development time that yields the highest net reproductive rate can be found:

$$\frac{dR_0}{d\kappa} = 0 \Rightarrow \kappa = \frac{c}{\sigma} \tag{7.21}$$

which is to say that optimum development time varies inversely with mortality rate and independently of fecundity. That is, the higher the mortality, the shorter the development time, and the earlier the maturation. This makes sense: if the mortality rate is high, copepods should develop fast to enhance their chance of becoming mature and reproducing, even though the reproductive rate may be low. If one assumes reasonable numbers for c (say 2) and σ (0.1 d^{-1} at 15°C, Hirst and Kiørboe 2002), then the prediction is that copepods should have a development time of about 20 d at 15°C, which they have (Huntley and Lopez 1992).

As mentioned, the above idea applies—and has been applied—more generally, and there is a huge literature on the topic (see, e.g., Stearns 1992). An interesting marine example is that of Atlantic cod off southern Labrador and eastern Newfoundland. Here elevated fishing mortality has apparently led to cod maturing earlier, thus suggesting fishing-induced evolution (Olsen et al. 2004). A similar case has been made for North Sea plaice (Grift et al. 2007).

7.9 INTERACTING POPULATIONS

Population interactions cover two main phenomena, competition and predator-prey (or parasite-host) interactions. Literally speaking, predator-prey interactions are not interactions between populations; it is the members of the populations that interact by eating one another. This has, of course, implications for the dynamics of the populations as it affects mortality, growth, and reproductive rates. Competition can better be considered as genuine population interaction, as it is typically the population of one species that reduces the availability of resources for individuals of the other (and vice versa). Below we consider some examples of predator-prey and competitive interactions between pelagic populations.

7.9.1 Predator-Prey Interactions

The population dynamic effects of predator-prey interactions have been described by the classical Lotka–Volterra equations (developed independently by Lotka and Volterra at the beginning of the last century).

Although these equations are gross simplifications of the real world, they provide a very good platform for examining predator-prey interactions in the pelagic, and they turn out to have surprisingly good predictive power. They are formulated as a set of coupled differential equations:

$$\frac{dC_{\text{prey}}}{dt} = \mu C_{\text{prey}} - \beta C_{\text{prey}} C_{\text{predator}}$$

$$\frac{dC_{\text{predator}}}{dt} = Y\beta C_{\text{prey}} C_{\text{predator}} - \sigma C_{\text{predator}}$$

(7.22)

What the equations say in words is this: for the prey (upper equation), the rate of change of prey density is equal to its growth rate (first term on right-hand side) minus the mortality from predation (second term). We know already what the latter is, namely the ingestion rate of the predator, which is the volumetric encounter rate (or clearance rate), β, multiplied by the concentrations of predators and prey (C_{prey} and C_{predator}). For the predator (lower equation), the rate of change in population size equals its growth rate (first term) minus its mortality rate (second term). We estimate the growth rate as the rate at which the predator eats prey ($\beta C_{\text{prey}} C_{\text{predator}}$) multiplied by the growth yield, Y. The latter is simply the inverse of the number of prey that the predator needs to eat in order to divide once (for a unicellular organism). So, the logic is simple, and the equations link readily to processes at the individual level that we have considered previously.

The properties of the equations are treated in many textbooks, to which I refer (e.g., Pielou 1969). The solution predicts coupled oscillations between predator and prey populations. Such oscillations, which accord with the predictions of the equations, have been demonstrated for, for example, populations of bacteria and heterotrophic flagellates in the ocean (the latter feed on the former) (Fenchel 1982). We can solve the equations for steady state by equating the derivatives with zero:

$$\hat{C}_{\text{prey}} = \frac{\sigma}{Y\beta}$$

$$\hat{C}_{\text{predator}} = \frac{\mu}{\beta}$$

(7.23)

The steady state is not stable; that is, even small deviations from steady state will lead to fluctuations. In that sense, the steady-state solution is of little practical interest. However, it can be shown that the average population sizes of the fluctuating populations are equal to the steady-state values (Pielou 1969). As an example, we can insert values for 1-µm pelagic bacteria (prey) and their 10-µm heterotrophic flagellate predators.

We have already estimated the flagellate clearance rates (β) for this interaction to be about 10^5 body volumes per hour (see section 3.6). We can estimate the growth yield by assuming that the flagellate has to eat bacteria corresponding to three times its own volume in order to divide. Because the volume of the prey bacteria is 1000 times less the flagellate, it follows that the flagellate should eat about 3000 bacteria per cell division; hence, $Y = 1/3000$. The mortality rate of the flagellates (σ) must be similar to their growth rate in the ocean, of order 1 d^{-1}. Realized average growth rates of pelagic bacteria in the ocean (μ) are also about 1 d^{-1}. Inserting these values yields bacterial concentrations of about 10^6 ml^{-1} and flagellate concentrations of about 10^3 ml^{-1}, both values that are typical for the marine environment.

The seasonal fluctuations in phytoplankton and zooplankton population sizes observed in seasonal, temperate seas (fig. 7.10) have traditionally been considered the result of a classical predator–prey fluctuation—the phytoplankton blooms in spring when zooplankton abundances are low, and the responding zooplankton populations then graze down the phytoplankton. We now know that the mesozooplankton typically eat only a small fraction of the spring phytoplankton bloom and, hence, that the spring bloom declines for other reasons (see chapter 8). Also, the time lag of one to several months between the peak of the phytoplankton and the peak of the copepods is inconsistent with the Lotka–Volterra model. It can be shown (Pielou 1969) that this time lag should be $0.5\pi(\mu\sigma)^{-0.5}$. Assuming a copepod mortality of 0.1 d^{-1} and a phytoplankton growth rate of 1 d^{-1}, this predicts a time lag of less than a week. The observed time lag is about one order of magnitude longer (fig. 7.10).

7.9.2 Attached Microbial Communities

We shall next consider an example of population interactions that draw on many of the mechanisms and examples we have examined in the previous chapters, namely that of the microbial communities that develop on suspended particles in the ocean, particularly marine snow. Pelagic microbial populations are often thought of as freely suspended, but in fact, a significant fraction of the populations may occur attached to particles. The estimates of the fraction of pelagic bacteria, for example, that are attached to particles are highly variable, from almost none to most (Simon et al. 2002). Although this may reflect the variation occurring in the ocean, it may also reflect difficulties in assessing the actual status (attached or "free") of the organisms. Microbial concentrations on, for example, marine snow aggregates are typically orders of magnitude higher than those in the ambient water (Alldredge et al. 1986, fig. 7.11). Often, but not always, the bacterial community attached to particles dif-

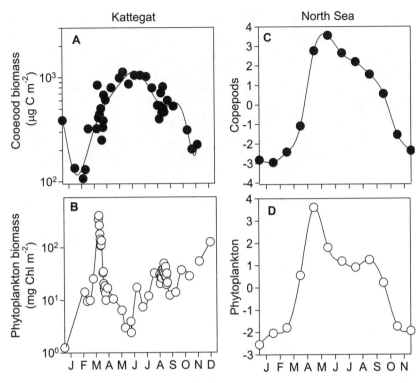

Fig. 7.10. Seasonal variation in the concentration of planktonic copepods in the Kattegat (Kiørboe and Nielsen 1994) and the North Sea (Colebrook 1979). The units for the North Sea data are standard deviations from the annual mean. Modified from Kiørboe (1998).

fers from those living in the ambient water, suggesting that some species of bacteria are particle specialists (Riemann and Winding 2001). The metabolic state of attached versus free microbes may also vary substantially, and the activity is typically much higher in attached than in free bacteria (Simon et al. 2002, Grossart et al. 2007). In any case, there is significant microbial activity associated with suspended particles (Azam and Long 2001). One implication of this activity is that sinking aggregates, for example, degrade in the water column rather than sediment to the sea floor. These processes, therefore, have significant implications for the structure of pelagic food webs (see chapter 8).

Suspended particles are colonized by a variety of microorganisms. Bacteria typically colonize first; then heterotrophic flagellates and ciliates follow. The flagellates feed on the attached bacteria, and the ciliates feed on both bacteria and flagellates. Eventually, if the particle is large

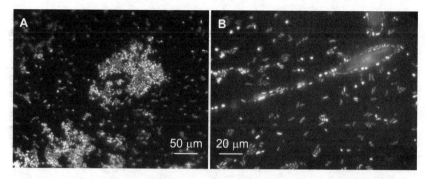

Fig. 7.11. Dense bacterial populations develop on the surface of suspended particles. In A, the attached bacteria have been stained with PicoGreen; in B with Dapi. Note in B the distribution of bacteria around the diatom (with chloroplasts appearing gray). This diatom, *Navicula*, slides over the surface of the particle, and bacteria gather in its wake, presumably because of locally elevated availability of organic material. Courtesy of Hans-Peter Grossart.

enough, small copepods may colonize, feeding on everybody else. Thus, complete and complex microbial communities may develop on suspended particles. There appears to be a characteristic scaling of the abundances of microbes attached to suspended particles, with microbial abundances increasing approximately with particle radius raised to the power 1.5. This applies to bacteria and heterotrophic flagellates as well as ciliates (Kiørboe 2003). These (steady-state) abundances are the result of a number of dynamic processes. Microorganisms colonize particles from the ambient water (cf. section 2.9), they may leave the particle again (detach), and they may both grow and become eaten while attached. For example, attached flagellates may graze on attached bacteria. If we initially consider only bacteria and heterotrophic flagellates, we can modify the Lotka–Volterra equations to take colonization and detachment into account in a description of the dynamics of attached microbial communities:

$$\frac{dB}{dt} = \beta_b' B_a + \mu B - \lambda_b B - \beta_{fb}'' FB$$

$$\frac{dF}{dt} = \beta_f' F_a + Y\beta_{fb}'' BF - \lambda_f F$$

(7.24)

where B and F are bacterial and flagellate densities on the particle (with dimensions of numbers per area), B_a and F_a are ambient bacterial and flagellate concentrations (dimensions numbers per volume), subscipts b and f refer to bacteria and flagellates, respectively, and the λs are specific detachment rates. There are several βs in the equations, and they, as usual, refer to encounter kernels for various processes (encounters between bacteria or flagellates in the ambient water with the particle, encounters between attached flagellates and their attached bacterial prey, β_{fb}). If we go through the bacterial equation first, then what the equation says is that the change in density of attached bacteria equals the colonization rate (first term on right side) plus the growth rate (second term) minus the detachment rate (third term) and minus the mortality from flagellate grazing (last term). The flagellate equation similarly says that the rate of change of density of attached flagellates equals the colonization rate of flagellates from the ambient water (first term) plus the growth from feeding on bacteria (second term) and minus the detachment rate (third term; this term can also be interpreted as a mortality rate or any gross loss term from the aggregate). You will note that the βs are marked with one or two primes; this is because they have been normalized to the surface area of the particle (to have dimensions LT^{-1}) or because, in the case of flagellate grazing, they have units L^2T^{-1}. (As an exercise: check for yourself that the dimensions of equation 7.24 are consistent.) The flagellates are clearing the surface of the particle for bacteria, not clearing the ambient water; hence, the different dimensions. The analytic solution to equation 7.24 is very complicated and, hence, difficult to examine generally, but we can simulate the development of the attached communities with realistic values of the different parameters. The predicted dynamics of the attached bacteria-flagellate community are very different from those predicted by the original Lotka–Volterra equations. Whereas the latter predicts coupled oscillations, equation 7.24 predicts development toward steady state (fig. 7.12).

We may modify equation 7.24 by adding yet another trophic level, ciliates colonizing from the ambient water and feeding on the attached flagellates. This would just be adding a third differential equation of the same form as the equation describing the flagellate dynamics. All the parameters in equation 7.24 (or its extension) can be measured or estimated from measurements (Kiørboe 2003). We have already talked about measuring particle colonization rates (section 2.9) and predator–prey encounter rates (above), but growth and detachment rates can also be measured. It turns out that many of these processes are density dependent. For example, growth rates of attached bacteria decline with increasing bacterial density, eventually because the diffusive transport of oxygen toward the particle may constrain bacterial growth at high densities (see section 3.5). Similarly, flagellates feeding on bacteria have a type II functional response (section 6.1) and thus become saturated at high bacterial densities. We can modify equation 7.24 to take density dependence into account. Again, all these functional relations can be

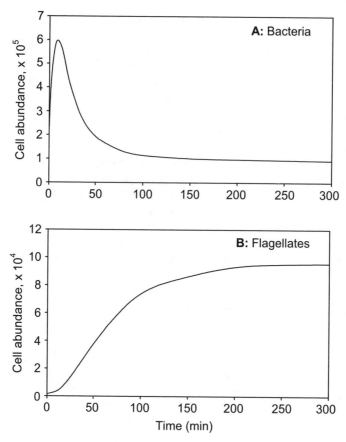

Fig. 7.12. Simulated development of densities of bacteria (A) and flagellates (B) on a suspended particle, using equation 7.24 and realistic parameter values.

quantified from experiments and behavioral observations, allowing one to predict the microbial dynamics on particles in laboratory experiments from a mechanistic and quantitative understanding of the component processes (fig. 7.13A). Evidently, both observations and predictions show that the microbial community rapidly reaches steady-state population densities.

The flow environment of real marine snow aggregates differs from that of artificial laboratory particles because the former sink through the water. This difference will increase colonization rates of microbes in the sea, but we can predict the increase from the Sherwood number (section 3.4). Again we can modify equation 7.24 accordingly. This modification allows us to predict steady-state microbial abundances on sinking

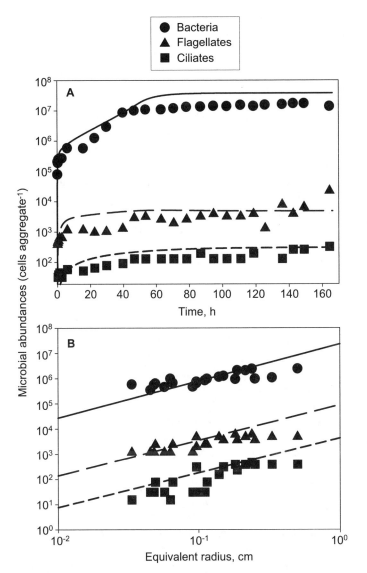

Fig. 7.13. Observed and predicted development of microbial populations on the surface of a suspended particle in a laboratory experiment (A) and observed and predicted steady-state densities of microbes on the surface of marine snow aggregates collected in the Øresund (B). Dots represent observed values, lines are predicted values (i.e., not regression lines). Laboratory data are from Kiørboe et al. (2003), field data from Grossart et al. (2003). Modified from Kiørboe (2003).

marine snow aggregates, and we may examine how these abundances depend on the sizes of particles by taking size-specific differences in sinking rate into account. Finally, we can compare the predicted scaling of abundances with those observed (fig. 7.13B), and we may note that what we observe is largely (within an order of magnitude) consistent with what we predict. The analysis is, of course, crude in the sense that we have treated bacteria, flagellates, and ciliates as homogeneous groups, although in fact they are diverse in terms of both taxonomy and functionality. Also, we have ignored the dynamics of the aggregates themselves as they form and degrade. There is thus room for improvement of the description. However, variation related to species-specific variations in function and population dynamics and from the dynamics of the particles becomes oppressed over the eight orders of magnitude variation in population abundances considered in figure 7.13. With these limitations, we are thus able to describe properties of the populations and their dynamics from a mechanistic and quantitative description of individual-level processes.

7.9.3 Phytoplankton Competing for Light

The final example we shall consider is competition for light between phytoplankton species. The phytoplankton model presented earlier in this chapter (section 7.3) can readily be changed to a multispecies model by writing species-specific population dynamics and growth equations (i.e., equations similar to 7.6 and 7.8), one set for each species, and by having the light attenuation in the water column add the contributions from all species (modify eq. 7.7) (Huisman et al. 1999a). Even without the multispecies extension of the model, one can predict that the species with the lowest critical light intensity will eventually win the competition because it is able to reduce the light intensity below the critical intensity for all other species—it will shade them all out. However, using parameters estimated in monoculture experiments, one can predict quite accurately the temporal development of multispecies population densities in experimental water columns (fig. 7.4C). Real-world situations are almost always more complicated than those considered in such simple models. For example, differences in sinking, buoyancy, and turbulent mixing will modify the loss terms and light climates experienced by the different species, thus modifying the outcome of the competition as dependent on, for example, the intensity of vertical mixing. However, because our model is mechanistic rather than just descriptive, it may readily be modified to take such factors into account and be used to predict the outcome of competition between species in nature (Huisman et al. 2004). Consider, for example, two species, one positively buoyant, the other sinking. In this situation it is not only the light sensitivity that determines the outcome of the competition but also the mixing regime: the buoyant species may win during periods of low turbulence, as it may form dense populations at the surface and thus shade out the other species. The sinking species, in contrast, may win during periods of stronger turbulent mixing if it is less light sensitive than the buoyant species. This may resemble the situation in the Baltic Sea, where dense surface populations of buoyant cyanobacteria develop in summer during periods of low-wind mixing intensity (Walsby et al. 1997). This simple mechanism may also account for coexistence of species rather than competitive exclusion in situations where the intensity of turbulent mixing varies, for example with the tides. Thus, Lauria et al. (1999) suggested that the coexistence of dinoflagellates and diatoms in a tidally influenced estuary was facilitated by variable tidal mixing: the dinoflagellates aggregate at the surface during slack periods but are mixed deeper in the water during ebb and flood currents, whereas diatom populations depend on periods of intense turbulence to remain suspended in the water column. Constant low or strong mixing would lead to one or the other group winning the competition.

7.10 From Individual to Population

We have considered several examples above that describe how population processes are the result of interactions occurring between and among individuals. The exercises illustrate that we can actually break down the population processes to individual interactions and, vice versa, we can predict population and community properties from insights into individual interactions. In two of the examples, we have been able to summarize the population processes resulting from quite complex interactions as simple scaling laws. They describe how population abundances scale with, respectively, the size of the organisms (copepods in the North Sea) or the size of the habitat (attached microbial communities).

One may ask why bother to go the long way from examining and quantifying all the individual component processes to arrive at simple scaling relations? It would have been much easier to derive in the first place, by simple statistical analysis (regression), the scaling laws from observed abundances. The main advantage of the mechanistic approach, in addition to satisfying our scientific curiosity, is that scaling laws—or other relationships—that are based on a causal understanding rather than just statistical descriptions have much higher predictive power. A mechanistic understanding of the component processes allows us to extrapolate relationships from one system or scenario to another with much higher confidence than if we were to extrapolate a statistical relation (a regression line, for example) beyond the environment from which it was derived. The latter, in fact, is rarely warranted. Eventually we may wish to describe and model complex systems, which requires that we lump functional groups and simplify complex processes. Mechanistic insights rather than statistical descriptions may prove particularly useful in such contexts.

Chapter Eight

STRUCTURE AND FUNCTION
OF PELAGIC FOOD WEBS

INDIVIDUAL- and population-level processes in the water column integrate to make up the pelagic food web. This chapter attempts to synthesize the insights achieved in previous chapters into a description of the structure and function of pelagic food webs. By *structure* we mean: which are the species that make up the pelagic food web, and how many are there of the different organisms? By *function* we mean: what are the roles of the different organisms and through which pathways and at what rates are energy and matter channeled? The basic scheme of energy and matter flow in the ocean is similar to that in terrestrial systems: inorganic carbon is reduced to organic compounds through photosynthesis and uptake of inorganic nutrients by green plants; subsequently, reduced carbon is oxidized by heterotrophic organisms in a series of steps whereby inorganic nutrients are simultaneously remineralized. The three-dimensional extension of the pelagic ecosystems, however, causes partial vertical separation of autotrophic and heterotrophic processes. Light attenuates vertically in the water column, and photosynthesis and uptake of inorganic nutrients by phytoplankton cells are therefore restricted to the photic zone, whereas oxidation and mineralization may occur throughout the water column. Although water often is the factor limiting plant biomass in terrestrial systems, the vertical separation of uptake and mineralization of mineral nutrients in pelagic systems often makes nutrients—typically phosphorus, nitrogen, and/or iron—the limiting factor for phytoplankton biomass in the ocean (Smetacek and Pollehne 1986). Therefore, physical processes that supply inorganic nutrients to the upper ocean, such as vertical mixing, become principal determinants of the structure and function of pelagic food webs.

The physical control of pelagic food-web structure occurs at a multitude of spatial and temporal scales. At the smallest micrometer to centimeter scale, physics governs the functioning of individual plankters, as we have seen in the previous chapter. At an intermediate meter scale, turbulent mixing moves plankters vertically in the water column and thus causes exposure to variable light intensity and at the same time brings inorganic nutrients into the photic zone. At larger scales, tides and currents transport plankton horizontally, and ocean circulation governs the oceanwide distribution of salt and refractory organic material. The physical forcing varies temporally, from short-term wind-mixing and upwelling events to seasonal variation in vertical water column structure and decadal variation in the intensity of the North Atlantic oscillation (NAO) and El Niño, for example. This chapter explores how physical forcing at small and intermediate scales, together with biological interactions, determines the structure and function of marine pelagic food webs.

It is mainly at the level of pelagic food webs that plankton ecology becomes directly relevant to applied problems and societal issues. Probably the most outstanding examples are the roles of pelagic food webs for fish production and fisheries in the ocean and for ocean carbon budgets and global climate. At the end of this chapter we shall discuss these issues.

8.1 Two Pathways in Pelagic Food Webs

Contemporary descriptions of pelagic food webs often distinguish two main pathways for the flow of matter and energy: the "classical" grazing food chain and the microbial food web (fig. 8.1). The grazing food chain was described first; hence its status as "classical" (see summary description by Steele 1974). According to the classical description, the majority of the phytoplankton production is consumed by mesozooplankton, mainly copepods, which in turn constitute the main food for planktivorous fish. Further, the main input of organic material to the seafloor was assumed to be sinking copepod fecal pellets. During the 1970s and 1980s, this description was modified, mainly because of the realization that pico- and nanophytoplankton (e.g., cyanobacteria) and heterotrophic microorganisms (bacteria and heterotrophic nano- and microflagellates) play much larger quantitative roles than previously believed. A new description of the pelagic food web, with inclusion of the "microbial" loop, was developed (Pomeroy 1974, Azam et al. 1983). According to this description, a significant fraction of the primary production consists of phytoplankters that are too small to be eaten by copepods and other mesozooplankters. They are instead eaten by small, heterotrophic flagellates that, in turn, are eaten by larger ciliates that may eventually be consumed by the mesozooplankters. Heterotrophic bacteria assimilate dissolved organic material that is lost from organisms or organic particles through leaking, cell lysis, or sloppy feeding and are themselves eaten by small flagellates; this "lost" organic material is in this way

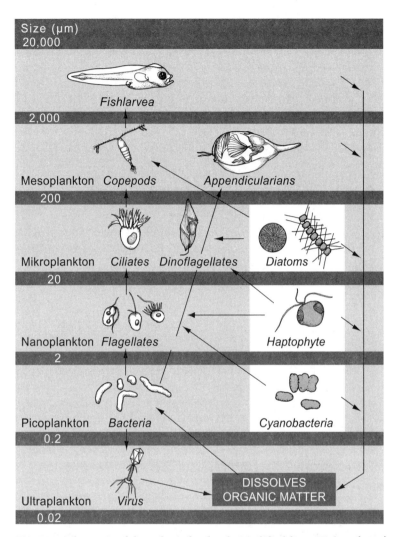

Fig. 8.1. Schematic of the pelagic food web. Modified from Nielsen (2005).

"looped" back to the food web. At about the same time it was realized that a significant fraction of the primary production at times may sediment directly—in the form of phytoplankton cells rather than fecal pellets—to the seafloor (e.g., Smetacek et al. 1978), and it was discovered that aggregation may facilitate this process (Billet et al. 1983, Alldredge and Silver 1988). More recently it has been discovered that viruses are abundant in the water column (Bergh et al. 1989) and of potential significance in regulating microbial abundances and community structure

(Fuhrman and Schwalbach 2003). Also, it has been found that "mixotrophy" is very common among certain quantitatively important protists, including both "algae" that are primarily phototrophic and "protozoa" that are primarily phagotrophic (Thingstad et al. 1996, Stoecker 1998). Mixotrophy is the phenomenon whereby an organism can function both as an autotroph (conduct photosynthesis) and a heterotroph (assimilate and utilize organic material), and their existence in significant numbers makes the traditional categorization of organisms doubtful.

The relative significance of the two pathways is closely related to turbulent mixing, water-column structure, nutrient regime, and cell size of the dominant phytoplankton primary producers (Smetacek and Pollehne 1986, Legendre and Le Fèvre 1989, Legendre 1990, Cushing 1989, Kiørboe 1993): small cells dominate in vertically stratified, oligotrophic waters and are mainly combusted in the microbial food web; the primary production here is mainly based on inorganic nutrients that are continuously being regenerated within the photic zone ("regenerated" production, Dugdale and Goering 1967). Large phytoplankton cells, in contrast, typically dominate in weakly stratified or well-mixed, eutrophic environments, where they sediment to the seafloor or are grazed by zooplankters and, thus, fuel production at higher trophic levels and support fisheries; the primary production here is based mainly on "new" nutrients transported by the environment to the photic zone ("new production," Dugdale and Goering 1967). Nutrient supply, mixing regime, and phytoplankton cell size thus appear to be correlated determinants of pelagic food web structure. In the following we shall try to reveal the causal, mechanistic relations.

8.2 LIGHT AND VERTICAL MIXING: CONDITIONS
   FOR PHYTOPLANKTON DEVELOPMENT

Light is prerequisite for phytoplankton growth, and because light attenuates with depth, phytoplankton cells need to remain suspended in the upper photic zone in order to maintain a population. Most phytoplankters are heavier than water and thus sink. In a water column without vertical turbulent mixing, all negatively buoyant phytoplankton will eventually sink to the sea floor, and it therefore requires a certain minimum turbulent intensity to maintain a growing pelagic phytoplankton population (Riley et al. 1949). (Note that nongrowing, sinking particles will end on the seafloor whether there is turbulence or not because turbulence does not alter the average sinking velocity; it is because the cells grow that turbulence makes a difference.) Because light attenuates with depth, mixing intensity similarly has to be below a critical value: if the phytoplankton population is mixed too deep in the water column, the average light intensity that it receives may be too low to allow net growth of the population, which will consequently go extinct. This is Sverdrup's (1953) "critical depth" concept. Huisman et al. (1999b, 2002) showed that viable populations may survive in environments even with deep mixing, provided the cells can outgrow sinking and mixing losses and that there is a "turbulence window" within which phytoplankton populations have positive net growth and thus can maintain a population. Turbulent mixing of a certain intensity thus is a prerequisite for pelagic phytoplankton populations to be retained in the photic zone and maintain a pelagic population.

8.3 BUDGETARY CONSTRAINTS: NUTRIENT INPUT AND SINKING FLUX

Turbulent mixing is also critical for phytoplankton populations in terms of nutrient supply. Consider an idealized ocean that is vertically stratified: an upper euphotic zone is separated from deeper waters by a thermocline. Primary production takes place in the upper mixed layer, where limiting inorganic nutrients become exhausted as they are built into biomass. A fraction of the photosynthetic products is combusted in the photic zone by heterotrophs, and the inorganic nutrients released during that process may fuel further "regenerated" primary production. Another fraction is lost from the upper layer through sinking of, for example, phytoplankton (aggregates) and zooplankton fecal pellets. This sinking flux drains limiting elements from the euphotic zone. At the same time, limiting nutrients are made available by turbulent diffusion and entrainment across the pycnocline from below the euphotic zone. These nutrients fuel "new" primary production. At steady state, or averaged over long times, the sinking flux of limiting elements has to exactly balance the net upward transport of limiting elements. Otherwise limiting elements in the form of dissolved nutrients or plankton biomass would build up infinitely or eventually be totally removed from the upper ocean. In the real ocean there may be other sources of limiting elements than transport from below, e.g., atmospheric deposition, but the same constraint will apply: the sinking flux has to balance the input of limiting elements in the long run. This balance often implies that export from the upper mixed layer in the form of sinking fluxes has to equal the rate of new production, which is therefore often termed "export production." In practice, export production is often estimated from measurements of the sinking flux.

This requirement for a stoichiometric mass balance implies that the planktonic community in the upper ocean is forced to adjust such that it

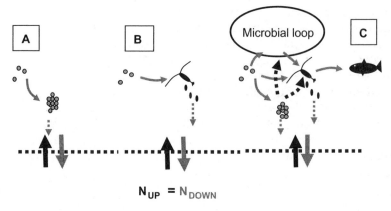

Fig. 8.2. Schematic exploring hypothetical food-web structures. See text for further explanation. Modified from Kiørboe (2001).

exactly produces the correct sinking flux. Because the delivery of new nutrients to the euphotic zone is almost entirely by physical processes (e.g., turbulent mixing), it follows that physical processes impose strong control on pelagic food-web structure and function.

Let us explore this constraint a bit further by next considering different idealized scenarios (fig. 8.2) in this idealized ocean that lead to several counterintuitive conclusions (Thingstad 1998, Kiørboe 2001). Assume first that the only source of sinking particles is phytoplankton (fig. 8.2A). Then the biomass of the phytoplankton community will increase until it produces the correct sinking flux. This will apply at steady state or, in a fluctuating environment, as a long-term average. Because sinking flux (Q_S) equals biomass concentration (C) times sinking velocity (v_S)

$$Q_s = C \cdot v_s \tag{8.1}$$

the steady-state (or long-term average) biomass is inversely proportional to the sinking velocity of the algae ($C = Q_S / v_S$) at a given input rate of limiting nutrients. Thus, if the flux is in the form of single cells that sink slowly, then the steady-state biomass concentration will be relatively high.

Conversely, if the algae coagulate into large and rapidly sinking aggregates, the biomass will be lower. The important point here is that the formation of aggregates increases the sinking velocity of particles but not the vertical flux of limiting elements. Aggregation thus has no influence on sinking fluxes (of limiting elements), but it instead reduces the pelagic biomass.

An alternative simplification is that all sinking is done by mesozooplankton fecal pellets (fig. 8.2B). In this case, mesozooplankton grazing, fecal-pellet production, and, hence, zooplankton biomass will adjust to produce the correct sinking flux. One important point here is that this balance will apply independent of the existence of a microbial loop and of the extent to which microbes remineralize phytoplankton production. In the real ocean, fisheries may partly "replace" sinking flux such that fisheries output plus sinking flux now has to balance the input rate of limiting elements, but the rate at which microbes combust phytoplankton production still has no influence on the magnitude of the fisheries. Thus, from this perspective, the microbial loop is neither a sink for nor a link to production at higher trophic levels (Ducklow et al. 1986).

A final and slightly more realistic scenario is that sinking consists of both phytoplankton (aggregates) and zooplankton fecal pellets (fig. 8.2C). Sinking of aggregates will reduce the need for fecal pellets and, hence, reduce the biomass of the zooplankton; and, conversely, sinking of fecal pellets will reduce the need for sinking algae and, hence, the steady-state concentration of phytoplankton. In general, the introduction of additional sinking agents (discarded mucus houses and feeding webs, exuvia, dead fish, etc.) will reduce the biomass of those components that produce the others. Conversely, processes that degrade the sinking pellets, aggregates, or other sinking particles, such as particle-colonizing microbes, and hence reduce particle sinking velocities would act to increase the steady-state pelagic biomass of phytoplankton and/or zooplankton. The stoichiometric constraint does not predict whether a community dominated by sinking of fecal pellets or by sinking of aggregates will develop. This trajectory is governed by other factors. Similarly, whether a pelagic community is dominated by small, slowly sinking particles at a relatively high biomass or large, rapidly sinking particles occurring at a relatively low biomass (cf. eq. 8.1)—or something in between—is not controlled by stoichiometry but by other factors.

The real ocean is, of course, more complicated than what has been described here. For example, one may argue that steady states of population sizes and fluxes are the exception rather than the rule in pelagic ecosystems, but the budgetary constraints—the mass balance of limiting elements—will still have to apply if averaged over long enough time periods. Also, there are more sources and sinks of limiting elements than considered above. For example, nitrogen may become available to primary production in the photic zone through direct fixation of atmospheric nitrogen by some phytoplankton species, and nitrogen may leave the photic zone by other processes than sinking and fisheries (e.g., denitrification and vertical migration). However, the overall conclusion is that the magnitude of the sinking flux is largely independent of the structure of the pelagic food web and is governed mainly by the physical processes that bring new nutrients into the photic zone. In contrast, the availability of inorganic nutrients in the photic zone and the nature (sinking velocity) of the sinking agents—a function of food-web structure—have direct influence on the overall pelagic biomass.

As discussed above, pelagic food-web structure appears to be closely related to the cell size of the dominant phytoplankton primary producers. Let us therefore continue our examination of food-web structure by considering the factors that decide which cell sizes win the competition for limiting nutrients and light in different environments.

8.4 CELL SIZE, WATER-COLUMN STRUCTURE, AND NUTRIENT AVAILABILITY: EMPIRICAL EVIDENCE

It has long been realized that small (<10 μm) phytoplankters dominate in oligotrophic and strongly stratified waters, whereas larger phytoplankters (net plankton) dominate in eutrophic, turbulent, and partially mixed water columns, particularly at spatiotemporal transitions between mixed and stratified waters, where they may form blooms (Margalef 1978, Malone 1980, Legendre 1981, Legendre et al. 1986, Cushing 1989, Chisholm 1992, Kiørboe 1993, Rodriguez et al. 2001). There are some exceptions to this pattern, particularly the occurrence of the floating mats of large-size *Rhizosolenia* diatoms in oligotrophic oceans that we shall return to below. However, there are numerous examples of the more typical pattern of blooms of net plankton—often diatoms—occurring at spatiotemporal transitions in water-column structure from mixed to stratified, and the subsequent replacement by nano- and picophytoplankton where and when the water column stabilizes. The classical example of such a succession is the seasonal development of the phytoplankton community during and subsequent to the vernal temperature stratification of the water column in temperate oceans. The net plankton blooms at the onset of stratification where nutrients are plentiful, and a nano- and picoplankton community develops subsequently during the oligotrophic, stratified summer period (e.g., Sournia et al. 1987; see also fig. 8.3A,B,F). Similar successions may be found on much smaller spatiotemporal scales, such as tidal fronts (Le Fèvre 1986, Richardson et al. 1985), shelf break fronts (Sharples et al. 2007), ephemeral wind-mixing events (Marra et al. 1990, Kiørboe and Nielsen 1990), and upwelling events (Peterson et al. 1988, Cermeñ̄o et al. 2006). In all such cases, enhanced phytoplankton growth conditions arise ephemerally: initial stratification retains phytoplankton in the photic zone, "new" inorganic

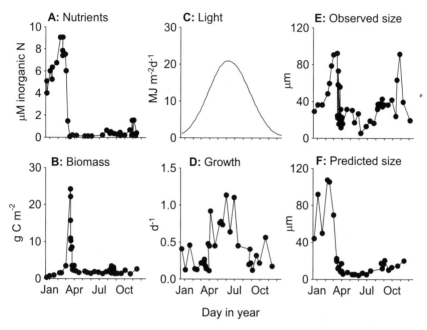

Fig. 8.3. Seasonal variation in observed nutrient availability (A) and light and phytoplankton characteristics (biomass, cell size, growth rate; B–E) in a temperate coastal ecosystem (Kattegat, Denmark). Observed cell sizes refer to the equivalent spherical diameter of the median (fiftieth percentile) of the volume-based size distribution (a measure of the maximum size) and is based on microscopic cell counts and cell size measurements; these data were collected by Dr. Gert Hansen, University of Copenhagen. Predicted maximum cell sizes (F) are computed from the ambient inorganic nitrogen concentration (A) and observed specific phytoplankton growth rate (D) using equation 2.25 and assuming an internal cell N content of 1.5 M. Panels A–D modified from Kiørboe (1996).

nutrients become available, and a net phytoplankton bloom develops. Similar blooms of large diatoms typically also result from experimental iron enrichment of iron-limited oceans (see section 8.12).

Another but related characteristic difference in the occurrence of small and large phytoplankton cells is that the abundance of the latter is much more variable than the abundance of the former (Malone 1980, Furuya and Marumo 1983, Malone et al. 1993). As a result, the relative contribution of picophytoplankton declines with increasing total phytoplankton biomass in large-scale comparisons, and the variation in phytoplankton biomass is governed almost entirely by variation in the biomass of the larger cells and is largely independent of the biomass of

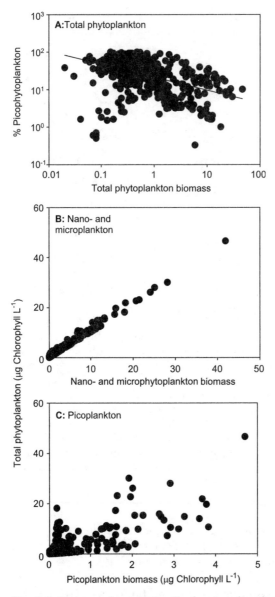

Fig. 8.4. Large-scale variation in the size distribution of phytoplankton. The relative contribution of picoplankton to total phytoplankton biomass as function of phytoplankton biomass (A), and the dependency of phytoplankton biomass on the biomass of nano- and microplankton (B) and picoplankton (C). Data from Agawin et al. (2000). Data were compiled by these authors from 38 studies conducted in a wide variety of habitats, from oligotrophic to eutrophic, from all major oceans.

the small cells (fig. 8.4). The variability of the nano- and net plankton abundance, quantified as the coefficient of variation, is more than five times the biomass of picoplankton in the large-scale comparison in figure 8.4. Thus, small picoplankton cells appear to occur at typical concentrations of about 10^4 cells ml^{-1}, independent of the temperature and nutrient status of the water (Fogg 1986). Small, suspended bacteria similarly occur at strikingly constant concentrations of about 10^6 cells ml^{-1}.

There are several possible explanations why large phytoplankton cells fluctuate widely in abundance and are dominant in ephemeral nutrient-rich habitats, whereas small cells occur at more constant abundances and dominate more stable environments. In the following we discuss the adaptive significance of cell size to nutrient uptake, turbulence, light, and predation.

8.5 CELL SIZE AND NUTRIENT UPTAKE

Nutrient limitation puts severe constraints on cell size. We have previously seen that diffusion-limited nutrient delivery to a phytoplankton cell scales with its size (radius) and, hence, that the specific nutrient uptake varies inversely with cell radius squared (section 2.5). In any environment, there is a maximum possible cell size, above which nutrient acquisition is insufficient to support significant growth, and in section 2.5 we derived an equation (eq. 2.25) that describes how this maximum possible cell size depends on the concentration of limiting nutrients and the phytoplankton growth rate. This equation may be applied to compute the seasonal variation in maximum possible cell size in a temperate, neritic ecosystem, where the seasonal variation in phytoplankton biomass and size distribution follows the textbook pattern described above (fig. 8.3). There is a striking similarity in observed and predicted cell size during spring and summer: in the course of the spring and during the peak of the spring bloom, both observed and predicted maximum cell size increases to about 100 μm diameter, whereupon it declines to about 10 μm, concurrent with depletion of inorganic nutrients in the surface layer. During late summer and autumn, the correspondence, however, is less convincing; this is the period when dinoflagellates dominate (Thomsen et al. 1992), many of which are mixotrophic and therefore less dependent on inorganic nutrients. Many dinoflagellates are also able to migrate below the pycnocline at night to acquire inorganic nutrients (Raven and Richardson 1984), further releasing them from nutrient limitation in the surface layer.

The similarity between observed and predicted cell size suggests that nutrient limitation is a main determinant of phytoplankton cell size in the ocean and that the size of dominant cells can be predicted from simple diffusion considerations. However, the similarity is perhaps more surprising than striking because even when nutrients occur in relatively high concentration, small cells are competitively superior to large cells in terms of nutrient uptake and possible growth rate. The more than ten-fold increase in cell size during the spring bloom corresponds to a factor of one hundred decrease in specific nutrient uptake rate and potential specific growth rate. It is therefore surprising that field populations of phytoplankton attain the maximum possible size. Why are they not out-competed by small, fast-growing cells? In chapter 3 it was shown that turbulence and sinking may enhance nutrient transport, and most so in large cells. However, the enhancement of nutrient uptake through sink-ing in a 100-μm diameter cell is only ~41 percent (table 3.1), which far from compensates the 10^2 times higher mass-specific potential nutrient uptake rate in a 10-μm cell (where the effect of sinking on nutrient up-take is negligible). Similarly, the positive effect of turbulence on nutrient uptake at realistic levels of turbulence is negligible relative to the nega-tive effects of size (employ eqs. 3.7–3.10). Therefore, turbulence and high nutrient availability allow the existence of large cells only in certain hab-itats but cannot explain why they are not outcompeted by small cells. The potentially most efficient compensation for large size in the compe-tition for nutrients is the diatom "trick" of inflating the cell size by inclu-sion of a large vacuole as it increases the mass-specific nutrient transport to the cell (section 2.7). However, at its best, the effect of inflated size is proportional to cell radius, whereas the negative effect of size varies with cell radius squared. This is thus also insufficient to account for the domi-nance of large cells in some habitats.

8.6 Cell Size, Turbulence, and Sinking

As noted above, phytoplankton depend on turbulent mixing to maintain a suspended population. Turbulent mixing at the scale of the depth of the upper mixed layer can be quantified as a turbulent diffusion coeffi-cient, D_{turb}, which has dimensions L^2T^{-1}. This is similar to ordinary mo-lecular diffusion coefficients, although typically turbulent diffusion coefficients are orders of magnitude higher than coefficients of molecu-lar diffusion, on the order of 0.1–100 cm^{2}s^{-1} in the upper ocean. The higher the sinking velocity of the cells in a population, the higher the intensity of the turbulence required to keep (some of) it suspended. It can simi-larly be argued that fast-growing cells require less turbulent mixing than slowly growing ones to overcome sinking losses. Thus, the minimal tur-bulence required to sustain a suspended population of growing cells is an increasing function of cell sinking velocity (u) and a decreasing function of the growth rate (μ). From dimensional analysis it follows that the minimum turbulent diffusion ($D_{\text{turb,min}}$) required to maintain a suspended population of growing cells is

$$D_{\text{turb,min}} = \frac{1}{4} \frac{u^2}{4\mu} \tag{8.2}$$

where the lead coefficient (1/4) comes from a more stringent and elaborate derivation of equation 8.2 (Riley et al. 1949, Huisman et al. 2002). It can be shown that this relationship holds true even in the presence of self-shading and the consequent decline in light and growth with depth, and that it provides a conservative estimate where background turbidity becomes significant (Huisman et al. 2002). In general, large cells sink faster than small cells (Smayda 1970). Stokes' law predicts that the sinking velocity of spherical cells increases with the square of their radius as long as the density difference between the cell and the ambient water is constant. In real algae, the density tends to decline with increasing cell size, and sinking velocity therefore rather increases with cell radius raised to a power of less than 2 (Jackson 1989). Still, though, it follows from equation 8.2 that the turbulent diffusion required to maintain a suspended population increases dramatically with cell size. This is further emphasized by the fact that phytoplankton growth rate declines with increasing cell size (Geider et al. 1986, Nielsen et al. 1996). Thus, for a 10-μm diameter cell sinking at 1 m d^{-1} and growing at a rate of 2 d^{-1} $D_{\text{turb,min}} = 10^{-2}$ cm^{2}s^{-1}, whereas for a 100μm diameter cell sinking at 20 m d^{-1} and growing at 0.5 d^{-1}, $D_{\text{turb,min}} = 20$ cm^{2}s^{-1}. In stratified water columns, the turbulent diffusion coefficient may be much less than the latter value and thus insufficient to keep the cells suspended. Therefore, turbulence is a prerequisite for the existence of populations of large, fast-sinking phytoplankters in the ocean, which is consistent with their absence from strongly stratified water columns. However, it still does not explain why large cells are not outcompeted by smaller ones, even in turbulent, ephemeral environments.

There are some notable exceptions to the general size–sinking velocity relation of Smayda (1970) and Jackson (1989) and to the general pattern of occurrence of large phytoplankters. Some cyanobacteria have gas vesicles and may thus be positively buoyant and form a surface population during periods of calm weather (Walsby et al. 1997). Many diatoms are capable of regulating their buoyancy through active modification of vacuolar size and ionic composition, and the largest species with the largest relative vacuole sizes may even become positively buoyant. The vacuole functions as a flotation device, and because the relative size of the vacuole in diatoms decreases with cell size, positive buoyancy is possible only in cells exceeding a certain critical minimum size of around 100 μm (Villareal 1988, Moore and Villareal 1996). Positive buoyancy is in particular found in the very large diatoms of the genus *Rhizosolenia* that may form dense mats at the surface of oligotrophic oceans; through buoyancy regulation they may migrate vertically to collect nutrients at depth and harvest light at the surface (Villareal et al. 1993).

8.7 CELL SIZE, TURBULENCE, AND LIGHT

The upper mixed layer is deeper and the intensity of the turbulent mixing typically higher in habitats dominated by large compared to those dominated by small cells, and the average intensity of the light is therefore less. Light requirements vary significantly between phytoplankton species with some species capable of growing at much lower light intensities than others. For example, cyanobacteria and dinoflagellates generally can grow at much lower light intensities than diatoms, but there seems to be no consistent relationship between size and light requirements for optimal growth (Raven and Richardson 1984).

The rate of photosynthesis in phytoplankton depends on the efficiency with which the cells harvest and utilize light and is directly dependent on three factors: how much chlorophyll the cell contains to absorb light, how well the chlorophyll can absorb the light, and how efficiently the absorbed light is utilized to reduce carbon to organic compounds. The first factor can be quantified relative to the carbon content of the cell (*Chl:C*, mg chlorophyll mg carbon^{-1}); the second factor can be quantified as an optical absorption cross-sectional area of the cell, for example, expressed relative to the chlorophyll content of the cell (A, m^2 mg chl^{-1}), and the last factor is the quantum efficiency of photosynthesis (ζ, mg C mol photon^{-1}). At a light intensity I (mol photon m^{-2}s^{-1}), the carbon specific gross rate of photosynthesis is (Geider et al. 1986)

$$P_c(I) = A\zeta(Chl:C)I \tag{8.3}$$

Each of the three factors that describe the light dependency of the gross photosynthetic rate may depend on cell size. However, both the chlorophyll-to-carbon ratio and the quantum efficiency appear to be independent of cell size, at least in diatoms (Geider et al. 1986). In contrast, measurements suggest that the absorption cross-section area declines with cell size (Geider et al. 1986, Finkel 2001). The simple, mechanistic explanation for this size dependency is self-shading inside the cell. An exception is large size attained by inflating the cell volume by means of a vacuole. This mechanism reduces the effect of self-shading and allows more efficient light harvesting (Raven 1997). Generally, however, size-dependent light harvesting favors small cells and does not help explain the dominance of large cells in turbulent environments, although the large vacuoles that are characteristic of many diatoms may partly offset the disadvantage of large size with respect to light harvesting.

8.8 WHY ARE NOT ALL PHYTOPLANKTERS SMALL?
THE SIGNIFICANCE OF PREDATION

The conclusion drawn from the previous sections is that small cells grow faster than large cells, independent of light, nutrient, and mixing conditions. Small cell size is, therefore, apparently competitively superior to large size under all circumstances, although the competitive pressure for small size may be somewhat relaxed in new, nutrient-rich, and turbulent habitats. Why, then, are not all phytoplankters in the ocean small?

The most likely reason that large phytoplankters can exist in the ocean is that they are capable of temporarily outgrowing their predators during events of enhanced growth conditions (Munk and Riley 1952, Geider et al. 1986, Kiørboe 1993), whereas small cells are controlled by their predators. Thus, size provides a partial refuge from predation in the plankton. Intuition as well as simple models demonstrate that this mechanism of "killing the winner" (the small cells) allows for the existence of competitive, inferior large phytoplankters (Thingstad 1998). There are two main reasons why large phytoplankters—and not small ones—can outgrow their predators. First, the relative density of predators decreases with size in pelagic food webs and, thus, relaxes predation pressure on large relative to small phytoplankton cells. It was originally suggested by Sheldon et al. (1972) that the size distribution of organisms in pelagic ecosystems is such that the average biomass in logarithmic size classes is approximately constant. Because prey size increases with predator size (prey:predator size on average about 0.1; fig. 6.6B), and because prey size spectra can be approximated by logarithmic distributions (chapter 6), this would imply constant predation pressure with increasing prey size in pelagic ecosystems. However, subsequent work has shown that pelagic biomass spectra actually have a negative slope, about -0.25 (Platt 1985), which implies decreasing relative predator density with increasing prey size. This effect is further emphasized by the declining prey:predator size with increasing prey size (fig. 6.6C). Secondly, small phytoplankton and cyanobacteria are grazed by unicellular protozoans that have generation times that are of similar order of magnitude as that of their prey cells. In contrast, large phytoplankters are typically grazed by large, multicellular zooplankters (copepods, euphausids) that have generation times that are orders of magnitude longer (weeks to years) than those of the prey (hours to days). This implies that a numerical response in predator density to changes in prey density is substantially lagged.

In an environment where enhanced phytoplankton growth conditions temporarily arise, both large and small phytoplankton cells will start to grow and increase in numbers. The small cells will rapidly be caught up by their protozoan predators, and their abundance will be controlled by predation. This predator control may explain the relative constancy of picoplankton populations in the ocean, a mechanism also suggested as being important for the regulation of population of pelagic bacteria (see chapter 7). The large phytoplankton cells, in contrast, will continue growing largely unutilized by their mesozooplankton grazers until all inorganic nutrients have been exhausted. They bloom! This is consistent with the very variable concentration of net phytoplankton in the ocean, and it explains why net plankton blooms at spatiotemporal discontinuities in water-column structure.

8.9 HYDRODYNAMIC CONTROL OF PELAGIC FOOD-WEB STRUCTURE: EXAMPLES

The physical-biological control of the phytoplankton size composition has implications for the structure of the pelagic food web and for the flow of energy and matter. Classical grazing-type food chains, with net plankton blooms and high mesozooplankton production and high sinking fluxes develop in new, ephemeral habitats during or following mixing events, whereas complex microbial food webs dominate in strongly stratified waters where the input of new nutrients is low and constant rather than pulsed. These patterns are particularly evident on ocean-wide and seasonal scales. Examples include the large, permanently stratified ocean gyres and the summer stratification in seasonal seas, where microbial food webs dominate, in contrast to, for example, the major upwelling regions and the vernal stratification in seasonal waters that are dominated by more "classical" food chains (Legendre and Le Fèvre 1989, Kiørboe 1993). The latter types of environments are also those that support the highest fisheries production in the world's oceans (Ryther 1969, Cushing 1989, Pauly and Christensen 1995), thus demonstrating that the hydrodynamic control of the phytoplankton production and size composition is, in fact, channeled to higher trophic levels. The sequence of events that occurs across spatiotemporal transitions from a mixed to a stratified water column is summarized in figure 8.5.

Similar patterns in food web structure may recur on much smaller spatiotemporal scales. Tidal fronts in the North Sea offer a good example (Kiørboe et al. 1988, Kiørboe 1993). In areas with strong tides, the tidal

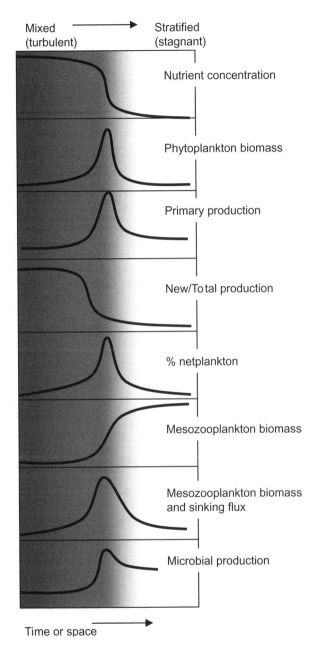

Fig. 8.5. Hydrodynamic control of pelagic food-web structure. Schematic summarizing the changes in pelagic food web characteristics across spatiotemporal variation in water-column structure and mixing regime. Modified from Kiørboe (1993).

energy may keep the water column permanently and almost entirely mixed close to the coast at depths less than a certain critical depth (Simpson and Hunter 1974). Further offshore and with increasing depth, tidal energy dissipates over a deeper water column, and hence, solar warming of the surface layer may overcome mixing, and the water column stratifies. A transect across one such tidal front shows a spatial change in water-column structure that resembles the temporal development in seasonal oceans during spring (fig. 8.6A). Offshore in the stratified water, a microbial-type food web may develop, whereas immediately inshore the frontal mixing may be too deep and intense to allow net growth of phytoplankton. Primarily because the tidal energy varies in a neap-spring cycle, the front moves back and forth, away from the coast during spring tide and toward the coast during neap tide. Thus, in a fortnightly cycle, mixed, nutrient-rich water becomes stratified, thus retaining phytoplankton in the photic zone with a plentiful supply of inorganic nutrients. As a consequence, a net plankton bloom develops (fig. 8.6B). This, in turn, leads to elevated productivity of the mesozooplankton (measured as rate of egg production in female copepods) (fig. 8.6C). The time and space scale of the event is such, however, that mesozooplankton biomass does not build up locally. Rather, mesozooplankton increases in an offshore direction and is highest in the stratified region; this is equivalent to the seasonal peak in mesozooplankton biomass in temperate seas, which occurs during summer stratification one to several months after the spring phytoplankton bloom (fig. 7.8). Despite the absence of a signal in mesozooplankton biomass, fish larvae accumulate at the front (fig 8.6D). Similar accumulations of larval fish at fronts have been observed repeatedly and for larvae of different species of fish (Munk et al. 1999), and the locally elevated mesozooplankton production may lead to elevated growth rate of the larval fish at frontal transitions (Munk 2007). The peak in netphytoplankton biomass at tidal and other types of fronts may finally lead to locally enhanced vertical flux of organic particles and, consequently, elevated biomass of benthic invertebrates (for an example, see Kiørboe 1996).

Other examples of smaller-scale spatiotemporal transitions in water-column structure with similar implications for pelagic food web structure as described above include other types of fronts, such as shelf break fronts (Fernandez et al. 1993, Sharples et al. 2007), shallow banks in deeper oceans (Nielsen et al. 1993, Richardson et al. 1998), wind mixing events (Kiørboe and Nielsen 1990, Frenette et al. 1996, Andersen and Prieur 2000, Nair et al. 1989), and short-term upwelling events (Peterson et al. 1988, Richardson et al. 2003, Zeldis et al. 2005, Ward et al. 2006). All such hydrographic events or phenomena enhance the upward transport of new nutrients ephemerally and/or locally, stimulate net plankton

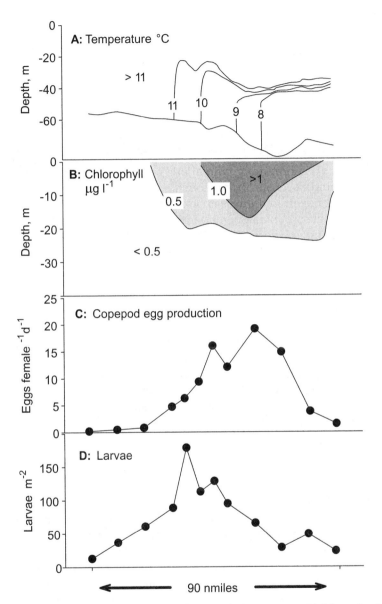

Fig. 8.6. Variation in food-web characteristics across a tidal front in the North Sea. Water-column structure as described by isopycnals (A), phytoplankton biomass (as concentration of chlorophyll) (B), copepod (*Acartia tonsa*) egg production rate (C), and concentration of herring larvae (D).

biomass accumulation and a classical type of food chain, and lead to elevated export in the form of sinking flux and/or in the form of elevated fish production and fisheries.

8.10 Species Diversity: The Paradox of the Plankton

One important aspect of food web structure is diversity: which are the species that make up the pelagic food web, and how many species can coexist in this apparently homogeneous environment? At least since Hutchinson (1961) formulated the "paradox of the plankton," researchers have tried to explain how a large number of plankton species, particularly phytoplankton, can coexist when they all appear to depend on the same few resources and are living in an apparently homogeneous environment. What puzzled Hutchinson was that competitive exclusion would allow only one species (per resource) to survive at steady state, as was demonstrated theoretically (section 7.9) and experimentally (fig. 7.4) in the previous chapter, yet dozens of phytoplankton species are typically found together in samples of a few milliliters. Hutchinson (1961) discussed several possible explanations of the paradox, including differential predation and temporal environmental variation.

The "killing the winner" principle discussed above demonstrates that differential predation will allow two and more competing species to coexist at steady state. Using simple steady-state models Thingstad (2005) was even able to reproduce the bell-shaped relationship between phytoplankton species diversity and resource availability that has been demonstrated in large-scale comparison of patterns in phytoplankton diversity (Irigoien et al. 2004). Thingstad (1998, 2000) extended the "killing the winner" idea to encompass pelagic bacteria. These also apparently live on a few common and limiting resources, yet many species may occur simultaneously. Thingstad has demonstrated that host specificity of viruses may account for coexistence among bacteria and has estimated that realistically up to a hundred competing bacterial species may coexist in a homogeneous environment and at steady state. Considerations of this type provide one solution to Hutchinson's paradox of the plankton.

However, there are several other processes that promote species diversity and coexistence, as also suggested by Hutchinson. First, the steady-state premise may be only rarely or at least not always fulfilled because of both environmental fluctuations and the nature of population interactions. Thus, Sommer (1983, 1984) demonstrated experimentally how several phytoplankton species could coexist when the input of a limiting nutrient was pulsed, whereas competitive exclusion would reduce the number of coexisting species to at most one per resource when the supply was constant. As discussed above, nutrient availability in the ocean may be pulsed on a multitude of temporal scales, thus potentially promoting phytoplankton diversity. Lack of steady state may also occur in stable environments. Resource–consumer interactions may lead to continuous population oscillations and even chaotic fluctuations, which may allow the coexistence of many species on few resources (e.g., Armstrong and McGehee 1980). This was demonstrated in particular for phytoplankton by Huisman and Weissing (1999) by a resource competition model: competition for three or more resources may lead to species oscillations and chaotic fluctuations in a stable environment, which allow coexistence among many species.

Patchiness may similarly promote species diversity by allowing contemporaneous nonequilibriums to develop (Richerson et al. 1970). Different patches dominated by different species may persist as long as growth rates exceed mixing rates but are destroyed frequently enough to prevent complete competitive exclusion within patches. Patchiness is well documented in the plankton, both on very small and on larger scales (e.g., Steele 1978, Franks and Jaffe 2001, Waters et al. 2003).

Patchiness in the plankton is often associated with spatial environmental heterogeneity, which may be significant in an apparently homogeneous ocean, as has been discussed at several instances in preceding chapters. Neighboring thin layers, for example, may be dominated by different phytoplankton species (fig. 8.7). Ephemeral solute plumes (fig. 3.7) may allow the survival of large, motile bacteria that have chemosensory capability (Blackburn et al. 1998) in an otherwise nutritionally dilute environment where the smaller bacteria are competitively superior. Suspended particles, including marine snow, may house bacterial "particle specialists" that at times are different from those species that dominate in the ambient water, particularly when particles are abundant during the end of a phytoplankton bloom (e.g., Riemann and Winding 2001). Finally, there may be many more different resources (substrates) for bacteria than just "organic solutes," thus allowing for a high species diversity. Substrate specificity of attached bacteria may lead to different algal species housing different bacterial communities (Grossart et al. 2005), and the endproduct of the activity of one bacterial type may be the substrate for another one. For example, one group of bacterial nitrifiers oxidizes ammonium to nitrite and another one nitrite to nitrate, but no organisms are known to be capable of doing both (Ward 2000). All these processes may prevent competitive exclusion and thus help promote diversity in the plankton.

Although there are thus several mechanisms that may account for species diversity in the plankton, it still remains an issue to document the diversity and to decide *how* diverse plankton communities are.

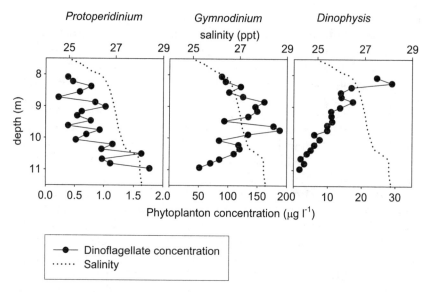

Fig. 8.7. Patchiness may allow apparent coexistence of competing species. Fine-scale vertical distribution of three species of dinoflagellates in the pycnocline of a temperate estuary. Samples were collected by a high-resolution water sampler that simultaneously collects 20 duplicate samples at 15-cm intervals in the vertical. Over 3 m in the vertical, the three species have overlapping but vertically segregated distributions. Data from Mouritsen and Richardson (2003).

Obviously, there are many undescribed species of everything from bacteria and phytoplankters to fish and probably even mammals, and our comprehension of plankton diversity is biased toward those species that are easy to collect and grow in the laboratory. It is well known, for example, that most species of bacteria cannot be cultured by known techniques. In order to document the diversity of protists by microscopy, these typically need to be concentrated from larger volumes of water by plankton nets, centrifugation, or preservation and sedimentation, and each concentration technique may reveal partly different species. Robust species, such as diatoms and thecate dinoflagellates, may survive any of these concentration techniques, but one wonders how many species are lost by all of the approaches. The same problem may also apply to larger plankton forms. Thus, the introduction of video-equipped remotely operated vehicles (ROVs) that can scan large volumes of water without destroying delicate organisms have led to the discovery of many new species of jelly plankton that are difficult to collect by traditional means (e.g., Hopcroft and Robinson 2005). Similarly, rare species are more likely to go undetected than common ones for obvious reasons.

Because microbes, in particular, are tremendously abundant (maybe 10^{30} individuals on earth, Curtis 2006), and because relative sample sizes are consequently extremely small, the sampling effort required to sample the rarest fraction of species is unrealistic. In (pelagic) microbes, this fraction of rare, never-to-be sampled species may be very high (Curtis and Sloan 2005).

In addition to—and partly because of—the practical problems of finding and identifying plankton species, there is an ongoing heated discussion of how many microbial species to expect in the ocean in the first place. One school (e.g., Fenchel 2005) argues that most microbes have almost cosmopolitan distributions, which prevents isolation and allopatric speciation. This leads to an expected relatively low species diversity, which is supported by the relatively few described protozoan species and "species" of bacteria (the latter mainly identified and characterized by molecular methods). The main argument is that the tremendous numbers of microbes in the ocean lead to a very high dispersion potential, such that "everything is everywhere" in the form of live cells or viable spores. The opponent school argues for a much greater degree of endemism among microbes with a consequently much higher diversity (e.g., Katz et al. 2005). This school would argue that because of the enormous numbers of microbes, insufficient sampling has revealed only a small fraction of the diversity (see above). Observations that pelagic bacteria have distinct biogeographies (i.e., are not everywhere, Pommier et al. 2005), that extractions and sequencing of genetic material from large-volume samples often yield an apparently high diversity of microbial species (e.g., López-Garcia et al. 2001, Venter et al. 2004), as well as the high frequency of "cryptic" microbial species (morphologically identical but genetically distinct) would argue for a high microbial diversity. The entire discussion is made difficult because the "species," particularly among asexual microbes, is not a well defined entity. Also, and maybe more important, genotypic variation, as revealed by rRNA sequences, may reflect accumulations of neutral mutations rather than functional diversity with the result the molecular approaches confuse rather than solve the issue of microbial diversity (Fenchel 2005). There is thus still little consensus regarding microbial diversity in the ocean, partly because genetic, physiological, and morphological groupings are difficult to relate to one another at this time (Dolan 2005).

8.11 Fisheries and Trophic Efficiency

Primary production, pelagic food-web structure, and fisheries production are obviously linked. In fact—and despite the potential complexity of pelagic food webs—there is a close correlation between rates of pelagic primary production and fisheries' yield (fig. 8.8A). The fisheries' magnitudes reported in figure 8.8 (from before 1980) can be assumed to be close to or slightly above the maximum sustainable yield in marine ecosystems because marine fish catches have been nearly constant since ~1980 despite increased effort (FAO 2005) and subsequently declining stocks (Pauly et al. 1998). The primary production rates in figure 8.8A refer to total production. However, it is only primary production based on new nutrients that allows a sustained fisheries' yield, as it is only new production that is available for export. One might therefore expect an even stronger relationship between fisheries' yield and new production than that observed with total production. There have been several attempts to synthesize relationships between new and total production (see summary by Dunne et al. 2005), all of which demonstrate that the relative contribution of new production increases with increasing productivity. This is expected in light of the above and corresponds to the observation that the variation in phytoplankton biomass is governed mainly by variation in the biomass of net plankton. If we use one such relationship (Wassmann 1990) that has the correct format for our purpose to convert estimates of total primary production to new production, we get the direct average relationship between fisheries' yield and new production (figure 8.8B). The indirect way of estimating new from total production does, of course, not strengthen the relationship, but it demonstrates that the slope is near 1 and therefore that fish production is nearly proportional to new primary production. (The same result is obtained if one uses the algorithm of Eppley and Peterson [1979] to convert total to new production.) The efficiency with which primary production is channeled to fish production decreases with primary production when total production is considered, but it is constant and independent of productivity if only new production is considered, and averages around 1 percent (figure 8.8C,D).

The proportionality between new production and fisheries emphasizes the realization above that the scope for export—including fisheries—is directly related to the rate at which new nutrients become available to primary production. Counter to most people's intuition, it is independent of the length of the food chain, the existence of a microbial loop, and of the extent to which microbes mineralize phytoplankton production. What rather determines the magnitude of a sustained fisheries' yield is the rate of new production.

Whether the fish production is dominated by pelagic or demersal fish may depend on the type of sinking agents that take care of the majority of the export. If the sinking flux is primarily caused by phytoplankton or phytoplankton aggregates, production of zooplankton and, hence, pe-

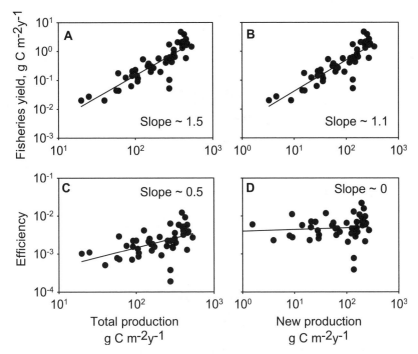

Fig. 8.8. Pelagic primary production and fisheries yield. Fisheries yield as a function of total primary production (A) and new primary production (B), and fisheries yield relative to production ("efficiency") as a function of total production (C) and new production (D). Data on fisheries yield and total primary production are from Nixon (1992). Estimates of total primary production have been converted to estimates of new production by using the algorithm of Wassmann (1990).

lagic fish would be relatively low, and demersal fisheries correspondingly high, compared to a situation where the sinking flux is to a larger extent a result of zooplankton fecal pellets. Can we make predictions about the relative significance of fecal pellets in the vertical material flux from our previous insights? As described above, the availability of new nutrients is often pulsed and gives rise to blooms of net phytoplankton. We have previously shown that phytoplankton aggregate formation by coagulation increases with the concentration of phytoplankton squared (section 4.3). This suggests that the higher the phytoplankton biomass during blooms, the larger the fraction of the phytoplankton will aggregate and settle to the seafloor, and the less is the need for zooplankton (fecal pellets) to produce the vertical flux. Thus, even though the large cells that typically dominate blooms fit the food size spectra of

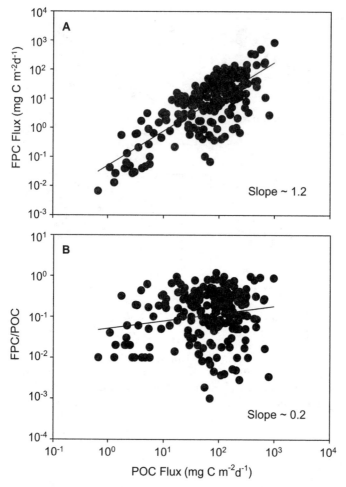

Fig. 8.9. The contribution of zooplankton fecal pellets to vertical particle flux. Vertical flux of mesozooplankton fecal-pellet carbon (FPC) as a function of total particulate organic carbon flux (POC) in the upper 200 m of the water column (A), and the relative contribution of fecal to total particulate organic carbon flux as a function of total flux (B). Data were compiled from studies covering a wide variety of habitats, including arctic, temperate, and tropical, as well as olig-otrophic and eutrophic oceans. Data are from Andreassen et al. (1996), Bauer-feind et al. (1997), Fowler et al. (1991), Fukuchi et al. (1993), González et al. (2000), Gowing et al. (2001), Huskin et al. (2004), Knauer et al. (1979), Miquel et al. (1994), Roy et al. (2000), Riebesell et al. (1995), Sampei et al. (2002), Silver and Gowing (1991), Taylor (1989), Urban-Rich et al. (1999), Urrère and Knauer (1981), Wexels Riser (1996), and Wexels Riser et al. (2002).

mesozooplankton well and even though the absolute magnitude of zooplankton production increases with injections of new nutrients, as demonstrated above, the relative magnitude of zooplankton production may
decrease with increasing productivity. Zooplankton fecal pellets may
therefore be expected to contribute a decreasing fraction of the sinking flux with increasing productivity or flux. Thus, we can make a prediction, but it seems to be wrong! The contribution of zooplankton fecal
material—although variable—appears to be independent of the magnitude of the sinking flux and, hence, of the rate of new primary production
when examined on a large scale (fig. 8.9A). This would suggest that the
relative significance of pelagic and demersal fish production should be
independent of the rate of new primary production.

On average, fecal material contributes about 10 percent of the sinking
flux (fig. 8.9B), which, in turn, suggests that mesozooplankton production corresponds to on the order of 10 percent of the new primary production (assuming that fecal-pellet production and growth are of similar
magnitudes, and disregarding mineralization of fecal pellets in the upper layer). With fish production corresponding to 1 percent of the primary production, this altogether brings memories of "pre-microbial-loop
times" when only the "classical" grazing food chain (phytoplankton-
zooplankton-fish) was considered and a trophic level efficiency of 10 percent was assumed. Without taking this comparison too far, it once again
demonstrates that the microbial loop can be considered neither a sink
nor a link and that the simple, old-fashioned description of a linear grazing food chain bears some resemblance to reality when only new production is considered.

8.12 FERTILIZING THE OCEAN—INCREASING THE FISHERY AND PREVENTING GLOBAL WARMING?

The observed relationship between new primary production and fisheries yield has led scientists and bioengineers to suggest that fertilizing the
ocean would lead to enhanced fisheries (Olsen 2002, Smetacek et al.
2004). Such effects have, in fact, been observed locally in coastal oceans
that have simultaneously experienced increased eutrophication and increased fisheries yield (Nielsen and Richardson 1996). Planned experiments on a large enough spatial and temporal scale to observe increased
harvesting potential at higher trophic levels have not, however, been
conducted, although visionary entrepreneurs (or fantasts) have taken
out patents on such techniques to increase ocean fish production.

Experimental fertilization of the ocean has, however, been conducted
several times now, but rather with the scope of examining the potential

of the so-called "biological pump" (sinking of particulate organic carbon) to sequester carbon and reduce atmospheric CO_2 content. Potentially, enhanced sedimentation of particulate organic carbon and subsequent burial in the ocean interior or in the sea floor may remove carbon dioxide from the atmosphere on a time scale long enough to have implications for global climate. All fertilization experiments conducted until now have used the micronutrient iron rather than nitrogen or phosphate macronutrients as fertilizer. The rationale is that the productivity in vast areas of the ocean—the so-called high-nitrogen–low-chlorophyll (HNLC) regions in the Pacific and Southern oceans—is limited by iron while the availability of macro nutrients is high. These areas cover about one-third of the ocean surface (Jickells et al. 2005).

The hypothesis that iron limits ocean productivity was suggested by Martin (1990), among others, and later it was verified directly by ocean iron addition experiments (Martin et al. 1994, Coale et al. 1996, Boyd et al. 2000). The purpose of the early fertilization experiments was mainly to test the iron limitation hypothesis and to explain why phytoplankton concentration is low in areas of the ocean where the concentration of macronutrients is high, a problem that had puzzled oceanographers for a long time. Alternative explanations for the existence of such areas, for example, that grazers keep phytoplankton populations permanently low (Frost 1991, Cullen 1991), had proven wrong or insufficient, but the idea that productivity was limited by iron availability was received with skepticism until eventually tested by ocean enrichment experiments. One of the observations that led Martin to the iron-limitation hypothesis was the negative correlation in ice cores between atmospheric dust and CO_2 concentrations over several hundred thousand years and the realization that an important source of iron in the open ocean is airborne dust transport (Martin 1990). This is also consistent with the notion that dust transport was high during glacial periods, when elevated iron availability increased the efficiency by which the biological pump transported CO_2 from the atmosphere to the deep ocean (Chisholm et al. 2001). Although the significance of dust transport and consequent changes in atmospheric CO_2 content has since been deemphasized (Jickells et al. 2005), the step from realizing that iron availability may regulate the downward flux of organic carbon in large parts of the ocean to suggesting that iron fertilization may be actively used to reduce atmospheric CO_2 content and reduce global warming is not far. The Southern Ocean would be the main target area for such a climate-engineering undertaking because here the availability and supply of macronutrients is high, and the transport of iron-containing dust is low. Model calculations suggest that with total depletion of macronutrients in the Southern Ocean, 1–2 Gt C could potentially be removed from the atmosphere each year, lowering anticipated ("business-as-usual scenario") atmospheric CO_2 increase by ~15 percent (Sarmiento and Orr 1991). Because the iron requirements are low (C:Fe molar ratio of phytoplankton ~10^5), the necessary amounts of iron and the implied costs would be manageable. A similar increase in new production and CO_2 drawdown based on nitrogen fertilization of N-starved regions of the ocean would require at least ~0.2 Gt of N per year (C:N ratio of phytoplankton ~6), which is similar to present-day production rate of nitrogen fertilizers (2005: 0.21 Gt yr^{-1} according to the statistics published by the International Fertilizer Industry Association, http://www.fertilizer.org/ifa/statistics/pit_public/pit_public_statistics.asp). This would, therefore, not be a feasible strategy to reduce atmospheric CO_2, also because energy (fossil fuels!) is required to produce nitrogen fertilizers.

Some of the first iron fertilization experiments in the Southern Ocean had difficulties demonstrating that the observed enhanced algal biomasses following iron enrichment in fact resulted in enhanced sinking flux of organic carbon (e.g., Boyd et al. 2000). This, of course, is a prerequisite for iron fertilization to reduce atmospheric CO_2 content. However, biomasses never reached concentrations where coagulation would become important for aggregate formation before bloom water was diluted by entrainment of surrounding waters, the patch dissipated, or the observations were terminated (Boyd et al. 2002, Jackson et al. 2005). However, unless there are other escape routes for iron that has been taken up by algae, budgetary considerations similar to those of section 8.3 dictate an eventual increase in sinking flux following iron-induced blooms. Subsequent longer-lasting experiments in the Southern Ocean have indeed demonstrated such elevated sinking fluxes (Coale et al. 2004), but the observed ratio of carbon flux enhancement to Fe enrichment is orders of magnitude less than the theoretical maximum of 10^5 (Buesseler et al. 2004).

The criticism of large-scale iron fertilization as a strategy to reduce the atmospheric CO_2 increase and global warming has, however, been severe. The main argument against the approach is that the modest benefit to the global carbon economy that such large-scale iron fertilization might have, 15 percent at best according to the above estimate, is not warranted by the risks of all the negative unknown and known possible side effects to the ecosystem (Chisholm et al. 2001). These side effects include possible development of anoxia at depth; this would favor microorganisms that produce methane, nitrous oxide, and other greenhouse gases that would counter CO_2 reduction in terms of global warming. It has also been argued that depletion of macronutrients in iron-fertilized regions would reduce new primary production downstream of the fertilization region, thus balancing local effects on CO_2 drawdown. Others have argued that these adverse effects are likely to be small, that ocean warming in the absence of iron fertilization will cause changes to the ecosystem anyway, and that it is unknown whether such changes are more or less significant than those caused by iron fertilization (Johnson and Karl 2002).

The global iron cycle and its impact on ocean biology, carbon budgets, and global climate are very complex and potentially allow large-scale feedback that may amplify or diminish future climate changes (Jickells et al. 2005). Present understanding of the processes is insufficient to accurately predict the outcome of iron fertilization on these large-scale processes, and an informed decision requires better data. The spatial and temporal scale of in situ iron enrichment experiments has increased over the past almost two decades, from covering about $60 \, km^2$ and lasting a few weeks in the first experiments conducted in the Equatorial Pacific (Martin et al. 1994), increasing to multiple subsequent enrichments resulting in patches covering $>1000 \, km^2$ and followed for 1 month in recent Southern Ocean experiments (Coale et al. 2004). Thus, apparently studies of the effects of iron enrichment have moved from an experimental scale to something that approaches a pilot scale. One can envisage that such a development will continue, with increased longevity and spatial extent of the fertilization experiments and with more or less continuous monitoring of relevant parameters both at the level of the individual (species composition, growth rates, etc.) and of the system (production, sinking fluxes, gas exchanges, etc.). This may, in fact, be the only way to achieve the necessary insights and appears to conform with the recommendations made by the British Department of the Environment to its government (Paper to the British Government Panel on Sustainable Development: Sequestration of Carbon Dioxide, July 1999, http://www.sd-commission.org.uk/panel-sd/position/co2/). The time scale of phytoplankton population dynamics is so short that the system is likely to return to preenrichment conditions very rapidly if enrichment ceases (Johnson and Karl 2002), thus allowing an escape route if adverse effects become significant.

Experiments in marine pelagic ecology are normally restricted to studiers of processes in bottles or mesocosms containing up to a few cubic meters of sea water, but such larger-scale ecosystem manipulations would offer unique opportunities to gain mechanistic insights into the functioning of pelagic food webs (Smetacek et al. 2004). One of the main aims of this book has been to demonstrate that insights in individual-level processes are critically important if one wants to understand how the biology of the ocean functions. The main challenge for such an approach is to directly establish the relationship between the functional ecology of the individuals and the function of the system and to test predictions about system functioning that are based on a mechanistic understanding at the level of the individual. Large-scale iron enrichment experiments might serve the double purpose of evaluating iron fertilization as a strategy to reduce atmospheric CO_2 and provide a test ground for process-based predictions of pelagic food web functioning. One can hope that scientists planning such experiments will take full advantage of these possibilities.

REFERENCES

Agawin, N. S. R., C. M. Duarte, and S. Agustí. 2000. Nutrient and temperature control of the contribution of picoplankton to phytoplankton biomass and production. *Limnol. Oceanogr.* 45:591–600.

Alldredge, A. L., J. J. Cole, and A. Caron. 1986. Production of heterotrophic bacteria inhabiting macroscopic organic aggregates (marine snow) from surface waters. *Limnol. Oceanogr.* 31:68–78.

Alldredge, A. L., and C. Gotschalk. 1988. In situ settling behavior of marine snow. *Limnol. Oceanogr.* 33:339–51.

Alldredge, A. L., and M. W. Silver. 1988. Characteristics, dynamics and significance of marine snow. *Prog. Oceanogr.* 20:42–82.

Andersen, V., and L. Prieur. 2000. One-month study in the open NW Mediterranean Sea (DYNAPROC experiment, May 1995): overview of the hydrobiogeochemical structures and effects of wind events. *Deep-Sea Res. I* 47:397–422.

Andreassen, I., E.-M. Nöthig, and P. Wassmann. 1996. Vertical particle flux on the shelf off northern Spitsbergen, Norway. *Mar. Ecol. Prog. Ser.* 137:215–28.

Andrews, J. C. 1983. Deformation of the active space in the low Reynolds-number feeding current of calanoid copepods. *Can. J. Fish. Aquat. Sci.* 40:1293–1302.

Armstrong, R. A., and R. McGehee. 1980. Competitive exclusion. *Am. Nat.* 115:151–70.

Azam, F., T. Fenchel, J. G. Field, et al. 1983. The ecological role of water-column microbes in the sea. *Mar. Ecol. Prog. Ser.* 10:257–63.

Azam, F., and R. A. Long. 2001. Oceanography: Sea snow microcosms. *Nature* 414:495–8.

Bagøien, E., and T. Kiørboe. 2005. Blind dating—mate finding in planktonic copepods. I. Tracking the pheromone trail of *Centropages typicus. Mar. Ecol. Prog. Ser.* 300:105–15.

Ban, S. 1994. Effect of temperature and food concentration on post-embryonic development, egg production and adult body size of calanoid copepod *Eurytemora affinis. J. Plankton Res.* 16:721–35.

Barbara, G. M., and J. G. Mitchell. 2003a. Marine bacterial organization around point-like sources of amino acids. *FEMS Microb. Ecol.* 43:99–109.

———. 2003b. Bacterial tracking of motile algae. *FEMS Microb. Ecol.* 44:79–87.

Bartumeus, F., F. Peters, S. Pueyo, et al. 2003. Helical Lévy walks: adjusting search statistics to resource availability in microzooplankton. *Proc. Natl. Acad. Sci. USA* 100:12771–5.

Bauerfeind, E., C. Garrity, C. Krumbholz, et al. 1997. Seasonal variability of sediment trap collections in the Northeast Water Polynya. 2. Biochemical and microscopic composition of sedimenting matter. *J. Mar. Syst.* 10:371–89.

Berg, H. C. 1993. *Random walks in biology.* Expanded ed. Princeton: Princeton University Press.

Berggreen, U., B. Hansen, and T. Kiørboe. 1988. Food size spectra, ingestion and growth of the copepod *Acartia tonsa*: implications for the determination of copepod production. *Mar. Biol.* 99:341–52.

Bergh, O., K. Y. Børsheim, G. Bratbak, et al. 1989. High abundance of viruses found in aquatic environments. *Nature* 340:467–8.

Billett, D. S. M., R. S. Lampitt, A. L. Rice, et al. 1983. Seasonal sedimentation of phytoplankton to the deep-sea benthos. *Nature* 302:520–2.

Blackburn, N., T. Fenchel, and J. Mitchell. 1998. Microscale nutrient patches in planktonic habitats shown by chemotactic bacteria. *Science* 282:2254–6.

Bollens, S. M., and B. W. Frost. 1991. Ovigerity, selective predation and variable diel vertical migration in *Euchaeta elongate* (Copepoda: Calanoida). *Oecologia (Berlin)* 87:155–62.

Boyd, P. W., G. A. Jackson, and A. Waite. 2002. Are mesoscale perturbation experiments in polar waters prone to physical artefacts? Evidence from algal aggregation modelling studies. *Geophys. Res. Lett.* 29:1541, doi:10.1029/2001GL014210.

Boyd, P. W., A. J. Watson, C. S. Law. et al. 2000. A mesoscale phytoplankton bloom in the polar Southern Ocean stimulated by iron fertilization. *Nature* 407:695–702.

Brown, K. A., and R. K. Zimmer. 2001. Controlled field release of a waterborne chemical signal stimulates planktonic larvae to settle. *Biol. Bull.* 200:87–91.

Buesseler, K. O., J. E. Andrews, S. M. Pike, et al. 2004. The effects of iron fertilization on carbon sequestration in the Southern Ocean. *Science* 304:414–7.

Buskey, E. J. 1995. Growth and bioluminescence of *Noctiluca scintillans* on varying algal diets. *J. Plankton Res.* 17:29–40.

———. 1997. Behavioral components of feeding selectivity of the heterotrophic dinoflagellate *Protoperidinium pellucidum*. *Mar. Ecol. Prog. Ser.* 153:77–89.

Butler, M. I., and C. W. Burns. 1991. Prey selectivity of *Piona exigua*, a planktonic water mite. *Oecologia* 86:210–22.

Caparroy, P., M. T. Perez, and F. Carlotti. 1998. Feeding behaviour of *Centropages typicus* in calm and turbulent conditions. *Mar. Ecol. Prog. Ser.* 168:109–18.

Catton, K. B., D. R. Webster, J. Brown, et al. 2007. Quantitative analysis of tethered and free-swimming copepod flow fields. *J. Exp. Biol.* 210:299–310.

Cermeñõ, P., E. Marañón, V. Pérez, et al. 2006. Phytoplankton size structure and primary production in a highly dynamic coastal ecosystem (Ría de Vigo, NW-Spain): Seasonal and short-time scale variability. *Est. Coast. Shelf Sci.* 67:251–66.

Chisholm, S. W. 1992. Phytoplankton size. In *primary productivity and biogeochemical cycles in the sea*, edited by P. G. Falkowski and A. D. Woodhead, 213–37. New York: Plenum Press.

Chisholm, S. W., P. G. Falkowski, and J. J. Cullen. 2001. Oceans—discrediting ocean fertilization. *Science* 294:309–20.

Clift, R., J. R. Grace, and M. E. Weber. 1978. *Bubbles, drops and particles*. New York: Academic Press.

Coale, K. H., K.S. Johnson, F. P. Chavez, et al. 2004. Southern Ocean Iron enrichment experiment: carbon cycling in high- and low-Si waters. *Science* 304:408–14.

Coale, K. H., Johnson, K. S., Fitzwater, S. E., et al. 1996. A massive phytoplankton bloom induced by an ecosystem-scale iron fertilization experiment in the equatorial Pacific Ocean. *Nature* 383:495–501.

Colebrook, J. M. 1979. Continuous plankton records—Seasonal cycles of phytoplankton and copepods in the North Atlantic Ocean and the North Sea. *Mar. Biol.* 51:23–32.

———. 1985. Continuous plankton records: overwintering and annual fluctuations in the abundance of zooplankton. *Mar. Biol.* 84:261–5.

Colin, S. P., J. H. Costello, and E. Klos. 2003. *In situ* swimming and feeding behavior of eight co-occurring hydromedusae. *Mar. Ecol. Prog. Ser.* 253:305–9.

Colin, S. P., and H. G. Dam. 2003. Effects of the toxic dinoflagellate *Alexandrium fundyense* on the copepod *Acartia hudsonica*: A test of the mechanisms that reduce ingestion rates. *Mar. Ecol. Prog. Ser.* 248:55–65.

———. 2004. Testing for resistance of pelagic marine copepods to a toxic dinoflagellate. *Evol. Ecol.* 18:355–77.

Costello, J. H., and S. P. Colin. 2002. Prey resource use by coexisting hydromedusae from Friday Harbor, Washington. *Limnol. Oceanogr.* 47:934–42.

Costello, J. H., R. Loftus, and R. Waggett. 1999. Influence of prey detection on capture success for the ctenophore *Mnemiopsis leidyi* feeding upon adult *Acartia tonsa* and *Oithona colcarva* copepods. *Mar. Ecol. Prog. Ser.* 191:207–16.

Costello, J. H., J. R. Strickler, C. Marrase, et al. 1990. Grazing in a turbulent environment—behavioral response of a calanoid copepod, *Centropages hamatus*. *Proc. Nat. Acad. Sci. USA* 87:1648–52.

Cowles, T. J., R. A. Desiderio, and M.-E. Carr. 1998. Small-scale planktonic structure: Persistence and trophic consequences. *Oceanography* 11:4–9.

Cowles, T. J., R. J. Olson, and S. W. Chisholm. 1988. Food selection by copepods: discrimination on the basis of food quality. *Mar. Biol.* 100:41–9.

Crimaldi, J. P., J. R. Hartford, and J. B. Weiss. 2006. Reaction enhancement of point sources due to vortex stirring. *Phys. Rev. E* 74:016307.

Cullen, J. H. 1991. Hypotheses to explain high-nutrient concentrations in the open sea. *Limnol. Oceanogr.* 36:1578–99.

Curtis, T. 2006. Microbial ecologists: it's time to "go large." *Nature Rev. Microbiol.* 4:488.

Curtis, T. P., and W. T. Sloan. 2005. Exploring microbial diversity—A vast below. *Science* 309:1331–3.

Cushing, D. H. 1989. A difference in structure between ecosystems in strongly stratified waters and those that are only weakly stratified. *J. Plankton Res.* 11:1–13.

Dam, H. G., and W. T. Peterson. 1988. The effect of temperature on the gut clearance rate-constant of planktonic copepods. *J. Exp. Mar. Biol. Ecol.* 123 (1):1–14.

Dekshenieks, M. M., P. L. Donaghay, J. M. Sullivan, et al. 2001. Temporal and spatial occurrence of thin phytoplankton layers in relation to physical processes. *Mar. Ecol. Prog. Ser.* 223:61–71.

DeMott, W. R. 1988. Discrimination between algae and artificial particles by freshwater and marine copepods. *Limnol. Oceanogr.* 33 (3):397–408.

———. 1989. Optimal foraging theory as a predictor of chemically mediated food selection by suspension-feeding copepods. *Limnol. Oceanogr.* 34 (1):140–54.

Denny, M. W. 1993. *Air and water. The biology and physics of life's media.* Princeton: Princeton University Press.

Doall, M. H., S. P. Colin, J. R. Strickler, et al. 1998. Locating a mate in 3D: The case of *Temora longicornis. Phil. Trans. R. Soc. B* 353:681–9.

Doall, M. H., J. R. Strickler, D. M. Fields, et al. 2002. Mapping the free-swimming attack volume of a planktonic copepod, *Euchaeta rimana. Mar. Biol.* 140:871–9.

Dolan, J. R. 2005. an introduction to the biogeography of aquatic microbes. *Aquat. Microb. Ecol.* 41:39–48.

Ducklow, H. W., Purdie, D. A., Williams, P. J. L., et al. 1986. Bacterioplankton—a sink for carbon in a coastal marine plankton community. *Science* 232:865–7.

Dugdale, R. C., and J. J. Goering. 1967. Uptake of new and regenerated forms of nitrogen in primary production. *Limnol. Oceanogr.* 12:196–206.

Dunne, J. D., R. A. Armstrong, A. Gnanadesikan, et al. 2005. Empirical and mechanistic models for the particle export ratio. *Global Biogeochemical Cycles* 19:GB4026, doi:1029/2004GB002390.

Dusenbery, D. B. 1996. Minimum size limit for useful locomotion by free-swimming microbes. *Proc. Natl. Acad. Sci. USA* 94:10949–54.

———. 1998. Fitness landscapes for effects of shape on chemotaxis and other behaviors of bacteria. *J. Bacteriol.* 180:5978–83.

Edwards, A. M., R. A. Phillips, N. W. Watkins, et al. 2007. Revisiting Lévy flight search patterns of wandering albatrosses, bumblebees and deer. *Nature* 449:1044–8.

Elliott, J. M. 2004. Prey switching in four species of carnivorous stoneflies. *Freshwater Biol.* 49:709–20.

———. 2006. Prey switching in *Rhyacophila dorsalis* (Trichoptera) alters with larval instar. *Freshwater Biol.* 51:913–24.

Eppley, R., and B. J. Peterson. 1979. Particulate organic matter flux and planktonic new production in the deep ocean. *Nature* 282:677–80.

FAO. 2005. *Review of the state of world marine fishery resources.* FAO Fisheries Technical Paper 457. Rome: FAO.

Fenchel, T. 1974. Intrinsic rate of natural increase—relationship with body size. *Oecologia* 14:317–26.

———. 1982. Ecology of heterotrophic microflagellates. I. Some important forms and their functional morphology. *Mar. Ecol. Prog. Ser.* 8:211–23.

———. 1984. Suspended marine bacteria as a food source. In *Flows of energy and materials in marine ecosystems,* ed. M. J. R. Fasham, 301–15, New York: Plenum Press.

———. 1986. The ecology of heterotrophic microflagellates. *Adv. Micob. Ecol.* 9:57–97.

———. 2001. Eppir si muove: Many water column bacteria are motile. *Aquat. Microb. Ecol.* 24:197–201.

———. 2005. Cosmopolitan microbes and their "cryptic" species. *Aquat. Microb. Ecol.* 41:49–54.

Fenchel, T., and N. Blackburn. 1999. Motile chemosensory behaviour of phagotrophic protists: Mechanisms for and efficiency in congregating at food patches. *Protist* 150:325–36.

Fernandez, E., J. Cabal, J. L. Acuna, et al. 1993. Plankton distribution across a slope-current induced front in the southern Bay of Biscay. *J. Plankton Res.* 15:619–41.

Fields, D. M., D. S. Shaeffer, and M. J. Weissburg. 2002. Mechanical and neural responses from the mechanosensory hairs on the antennule of *Gaussia princes*. *Mar. Ecol. Prog. Ser.* 227:173–86.

Fields, D. M., and J. Yen. 1997. The escape behavior of marine copepods in response to a quantifiable fluid mechanical disturbance. *J. Plankton Res.* 19:1289–304.

Fiksen, Ø. 1997. Allocation patterns and diel vertical migration: Modelling the optimal Daphnia. *Ecology* 78:1446–56.

Finkel, Z. V. 2001. Light absorption and size scaling of light-limited metabolism in marine diatoms. *Limnol. Oceanogr.* 46:86–94.

Fogg, G. E. 1986. Picoplankton. *Proc. R. Soc. Lond. B.* 228:1–30.

Fowler, S. W., and G. A. Knauer. 1986. Role of large particles in the transport of elements and organic compounds through the oceanic water column. *Prog. Oceanogr.* 16:147–94.

Fowler, S. W., L. F. Small, and J. La Rosa. 1991. Seasonal particulate carbon flux in the coastal northwestern Mediterranean Sea, and the role of zooplankton fecal matter. *Oceanologica Acta* 14:77–85.

Franks, P.J.S., and J. S. Jaffe. 2001. Microscale distributions of phytoplankton: initial results from a two-dimensional imaging fluorometer, OSST. *Mar. Ecol. Prog. Ser.* 220:59–72.

Frenette, J. J, W. F. Vincent, L. Legendre, et al. 1996. Size-dependent phytoplankton responses to atmospheric forcing in Lake Biwa. *J. Plankton Res.* 18:371–91.

Frost, B. W. 1975. Threshold feeding behaviour in *Calanus pacificus*. *Limnol. Oceanogr.* 20:263–6.

———. 1991. The role of grazing in nutrient-rich areas of the open sea. *Limnol. Oceanogr.* 36:1616–30.

Frost, B. W., and S. M. Bollens. 1992. Variability of diel vertical migration in the marine planktonic copepod *Pseudocalanus newmani* in relation to its predators. *Can. J. Fish. Aquat. Sci.* 49:1137–41.

Fuhrman, J. A., and R. L. Ferguson. 1986. Nanomolar concentrations and rapid turnover of dissolved free amino acids in seawater: Agreement between chemical and microbiological measurements. *Mar. Ecol. Prog. Ser.* 33:237–42.

Fuhrman, J. A., and M. Schwalbach. 2003. Viral influence on aquatic bacterial communities. *Biol. Bull.* 204:192–5.

Fukuchi, M., H. Sasaki, H. Hattori, O. Matsuda, A. Tanimura, N. Handa, and C. P. McRoy. 1993. Temporal variability of particle flux in the northern Bering Sea. *Cont. Shelf Res.* 13:693–704.

Furuya, K., and R. Marumo. 1983. Size distribution of phytoplankton in the Western Pacific Ocean and adjacent waters in summer. *Bull. Plankton Soc. Japan* 30:31–2.

Gascoigne, J., and R. N. Lipicus. 2004. Allee effects in marine systems. *Mar. Ecol. Prog. Ser.* 269:49–59.

Geider, R. J., T. Platt, and J. A. Raven. 1986. Size dependence of growth and photosynthesis in diatoms. *Mar. Ecol. Prog. Ser.* 30:93–104.

Gerritsen, J. 1980. Sex and parthenogenesis in sparse populations. *Am. Nat.* 115:718–42.

Gill, C. W., and S. A. Poulet. 1988. Responses of copepods to dissolved free amino acids. *Mar. Ecol. Prog. Ser.* 43:269–76.

Gismervik, I., and T. Andersen. 1997. Prey switching by *Acartia clausi*: Experimental evidence and implications of intraguild predation assessed by a model. *Mar. Ecol. Prog. Ser.* 157:247–59.

González, H. E., V. C. Ortiz, and M. Sobarzo. 2000. The role of faecal material in the particulate organic carbon flux in the northern Humboldt Current, Chile (23°S), before and during the 1997–1998 El Niño. *J. Plankton Res.* 22:499–529.

González, J. M., E. B. Sherr, B. F. Sherr. 1993. Differential feeding by marine flagellates on growing versus starving, and on motile versus nonmotile, bacterial prey. *Mar. Ecol. Prog. Ser.* 102:257–67.

González, J. M., and C. A. Suttle. 1993. Grazing by marine nanoflagellates on viruses and virus-sized particles: ingestion and digestion. *Mar. Ecol. Prog. Ser.* 94:1–10.

Gowing, M. M., D. L. Garrison, H. B. Kunze, et al. 2001. Biological components of Ross Sea short-term particle fluxes in the austral summer of 1995–1996. *Deep-Sea Res. I.* 48:2645–71.

Granata, T. C., and T. D. Dickey. 1991. The fluid mechanics of copepod feeding in a turbulent flow—a theoretical approach. *Prog. Oceanogr.* 26:243–61.

Grift, R. E., M. Heino, A. D. Rijnsdorp, et al. 2007. Three-dimensional maturation reaction norms for North Sea plaice. *Mar. Ecol. Prog. Ser.* 334:213–24.

Grossart, H. P., S. Hietanen, and H. Ploug. 2003. Microbial dynamics on diatom aggregates in Øresund, Denmark. *Mar. Ecol. Prog. Ser.* 249:69–78.

Grossart, H. P., F. Levold, M. Allgaier, et al. 2005. Marine diatom species harbour distinct bacterial communities. *Env. Microb.* 7:860–73.

Grossart, H. P., K. W. Tang, T. Kiørboe, et al. 2007. Comparison of cell-specific activity between free-living and attached bacteria using isolates and natural assemblages. *FEMS Microbiol. Lett.* 266:194–200.

Hairston, N. G., R. A. Van Brunt, and C. M. Kearns. 1995. Age and survivorship of diapausing eggs in a sediment egg bank. *Ecology* 76:1706–11.

Hamm, C. E., R. Merkel, O. Springer, et al. 2003. Architecture and material properties of diatom shells provide effective mechanical properties. *Nature* 42:841–3.

Hamm, C. E., D. A. Simonsen, R. Merkel, et al. 1999. Colonies of *Phaeocystis globosa* are protected by a thin but tough skin. *Mar. Ecol. Prog. Ser.* 187:101–11.

Hamner, P., and W. M. Hamner. 1977. Chemosensory tracking of scent trails by planktonic shrimp *Acetes Sibogae Australis*. *Science* 195:886–8.

Hansen, B., P. K. Bjørnsen, and P. J. Hansen. 1994. The size ratio between planktonic predators and their prey. *Limnol. Oceanogr.* 39:395–403.

Hansen, J., U. Timm, and T. Kiørboe. 1995. Adaptive significance of phytoplankton stickiness with emphasis on the diatom *Skeletonema costatum*. *Mar. Biol.* 123:667–76.

Hansen, J.L.S., and A. B. Josefson. 2003. Accumulation of algal pigments and live planktonic diatoms in aphotic sediments during the spring bloom in the transition zone of the North and Baltic Seas. *Mar. Ecol. Prog. Ser.* 248:41–54.

———. 2004. Ingestion by deposit-feeding macro-zoobenthos in the aphotic zone does not affect the pool of live pelagic diatoms in the sediment. *J. Exp. Mar. Biol. Ecol.* 308:59–84.

Hansen, P. J. 2002. Effect of high pH on the growth and survival of marine phytoplankton: implications for species succession. *Aquat. Microb. Ecol.* 28:279–88.

Hansen, P. J., P. K. Bjørnsen, and B. W. Hansen. 1997. Zooplankton grazing and growth: Scaling within the 2–2000-μm body size range. *Limnol. Oceanogr.* 42:687–704.

Hansen, P. J., and A. J. Calado. 1999. Phagotrophic mechanisms and prey selection in free-living dinoflagellates. *J. Eukaryotic Microbiol.* 45:382–89.

Hansson, L. J., and T. Kiørboe. 2006a. Effects of large gut volume in gelatinous zooplankton: Ingestion rate, bolus production, and food patch utilization by the jellyfish *Sarsia tubulosa*. *J. Plankton Res.* 28:1–6.

———. 2006b. Prey-specific encounter rates and handling efficiencies as causes of prey selectivity in ambush feeding hydromedusae. *Limnol. Oceanogr.* 51:1849–58.

Heuch, P. A., M. H. Doall, and J. Yen. 2007. Water flow around a fish mimic attracts a parasitic and deters a planktonic copepod. *J. Plankton Res.* 29 (Suppl. 1):i3–i16.

Hill, P. S. 1992. Reconciling aggregation theory with observed vertical fluxes following phytoplankton blooms. *J. Geophys. Res.* 97:2295–308.

Hill, P. S., A.R.M. Nowell, and P. A. Jumars. 1992. Encounter rate by turbulent shear of particles similar in diameter to the Kolmogorov scale. *J. Mar. Res.* 50:643–68.

Hirche, H. J. 1996. Diapause in the marine copepod, *Calanus finmarchicus*—A review. *Ophelia* 44:129–43.

Hirst, A. G., and T. Kiørboe. 2002. Mortality of marine planktonic copepods: global rates and patterns. *Mar. Ecol. Prog. Ser.* 230:195–209.

Holling, C. S. 1959a. The components of predation as revealed by a study of small mammal predation on the European pine sawfly. *Canad. Entomol.* 91:293–320.

———. 1959b. Some characteristics of simple types predation and parasitism. *Canad. Entomol.* 91: 385–98.

Hopcroft, R. R., and B. H. Robison. 2005. New mesopelagic larvaceans in the genus *Fritillaria* from Monterey Bay, California. *J. Mar. Biol. Ass. UK.* 85:665–78.

Huisman. J. 1999. Population dynamics of light-limited phytoplankton: Microcosm experiments. *Ecology* 80:202–10.

Huisman, J., M. Arrayás, U. Ebert, and B. Sommeijer. 2002. How do sinking phytoplankton species manage to persist. *Am Nat.* 159:245–54.

Huisman, J., R. R. Jonker, C. Zonneveld, et al. 1999a. Competition for light between phytoplankton species: Experimental tests of mechanistic theory. *Ecology* 80:211–22.

Huisman, J., P. van Oostveen, and F. J. Weissing. 1999b. Critical depth and critical turbulence: Two mechanisms for the development of phytoplankton blooms. *Limnol. Oceanogr.* 44:1781–87.

Huisman, J., J. Sharples, J. M. Stroom, P. M. Visser, W.E.A. Kardinall, J.M.H. Verspagen, and B. Sommeijer. 2004. Changes in turbulent mixing shift competition for light between phytoplankton species. *Ecology* 85:2960–70.

Huisman J., and F. J. Weissing. 1999. Biodiversity of plankton by species oscillations and chaos. *Nature* 402:407–10.

Hunter, J. R., and G. L. Thomas. 1974. Effect of prey distribution and prey density on the searching and feeding behaviour of larval anchovy *Engraulis mordax* Girard. In *The early life history of fish*, ed. J.H.S. Blaxter, 559–74. Berlin: Springer.

Huntley, M. E., and M.D.G. Lopez. 1992. Temperature-dependent production of marine copepods: A global synthesis. *Am. Nat.* 140:201–42.

Huntley, M. E., and M. Zhou. 2004. Influence of animals on turbulence in the sea. *Mar. Ecol. Prog. Ser.* 273:65–79.

Huskin, I., L. Viesca, and R. Anadón. 2004. Particle flux in the Subtropical Atlantic near the Azores: Influence of mesozooplankton. *J. Plankton Res.* 26:403–15.

Hutchinson, G. E. 1951. Copepodology for the ornithologist. *Ecology* 32:571–7.

———. 1961. The paradox of the plankton. *Am. Nat.* 95:137–45.

Incze, L. S., D. Hebert, N. Wolff, N. Oakey, and D. Dye. 2001. Changes in copepod distributions associated with increased turbulence from wind stress. *Mar. Ecol. Prog. Ser.* 213:229–40.

Irigoien, X., J. Huisman, and R. P. Harris. 2004. Global biodiversity patterns of marine phytoplankton and zooplankton. *Nature* 429:863–67.

Jackson, G. A. 1980. Phytoplankton growth rate and zooplankton. *Nature* 284:439–41.

———. 1987. Physical and chemical properties of aquatic environments. In *Ecology of microbial communities,* ed. M. Fletcher, T.R.G. Gray, and J. G. Jones, 213–33. Cambridge, UK: Cambridge University Press.

———. 1989. Simulation of bacteria attraction and adhesion to falling particle in an aquatic environment. *Limnol. Oceanogr.* 34:514–30.

———. 1990. A model of the formation of marine algal flocs by physical coagulation processes. *Deep-Sea Res.* 37:1197–211.

Jackson, G. A., and T. Kiørboe. 2004. Zooplankton use of chemodetection to find and eat particles. *Mar. Ecol. Prog. Ser.* 269:153–62.

Jackson, G. A., and S. E. Lochmann, 1992. Effect of coagulation on nutrient and light limitation of an algal bloom. *Limnol. Oceanogr.* 37:77–89.

Jackson, G. A., A. M. Waite, and P. W. Boyd. 2005. Role of algal aggregation in vertical carbon export during SOIREE and in other low biomass environments. *Geophys. Res. Lett.* 32: L13607, doi:10.1029/2005GL023180.

Jacobsen, D. M., and D. M. Anderson. 1986. Thecate heterotrophic dinoflagellates: Feeding behaviour and mechanisms. *J. Phycol.* 22:249–58.

Jakobsen, H. H. 2001. Escape response of planktonic protists to fluid mechanical signals. *Mar. Ecol. Prog. Ser.* 214:67–78.

———. 2002. Escape of protists in predator-generated feeding currents. *Aquat. Microb. Ecol.* 26:271–81.

Jakobsen, H. H., L. M. Everett, and S. L. Strom. 2006. Hydromechanical signalling between the ciliate *Mesodinium pulex* and motile protist prey. *Aquat. Microb. Ecol.* 44:197–206.

Jakobsen, H. H., and P. J. Hansen. 1997. Prey size selection, grazing and growth response of the small heterotrophic dinoflagellate *Gymnodinium* sp. and the ciliate *Balanion comatum*—a comparative study. *Mar. Ecol. Prog. Ser.* 158:75–86.

Jenkinson, I. R. 1995. A review of 2 recent predation rate models—The dome-shaped relationship between feeding rate and shear rate appears universal. *ICES J. Mar. Sci.* 52:605–10.

Jiang, H. S., and Osborn, T. R. 2004. Hydrodynamics of copepods: A review. *Surv. Geophys.* 25:339–70.

Jickells, T. D., Z. S. An, K. K. Andersen, et al. 2005. Global iron connections between desert dust, ocean biogeochemistry, and climate. *Science* 308:67–71.

Jiménez, J. 1997. Oceanic turbulence on the millimetre scales. *Scientia Marina* 61(Suppl. 1):47–56.

Johansen, J. E., J. Pinhassi, N. Blackburn, et al. 2002. Variability in motility characteristics among marine bacteria. *Aquat. Microb. Ecol.* 28:229–37.

Johnson, D. S., and D. M. Karl. 2002. Is ocean fertilization credible and creditable? *Science* 296:467–68.

Jonsson, P. R., and P. Tiselius. 1990. Feeding behaviour, prey detection and capture efficiency of the copepod *Acartia tonsa* feeding on planktonic ciliates. *Mar. Ecol. Prog. Ser.* 60:35–44.

Jumars, P. A. 1993. *Concepts in Biological Oceanography: An Interdisciplinary Primer.* Oxford: Oxford University Press.

Jürgens, K., and W. R. DeMott. 1995. Behavioral flexibility in prey selection by bacterivorous nanoflagellates. *Limnol. Oceanogr.* 40:1503–7.

Karp-Boss, L., E. Boss, and P. A. Jumars. 1996. Nutrient fluxes to planktonic osmotrophs in the presence of fluid motion. *Oceanog. Mar. Biol. Ann. Rev.* 34:71–107.

Katajisto, T. 1996. Copepod eggs survive a decade in the sediments of the Baltic sea. *Hydrobiologia* 320:153–59.

Katajisto, T., M. Viitasalo, and M. Koski. 1998. Seasonal occurrence and hatching of calanoid eggs in sediments of the northern Baltic Sea. *Mar. Ecol. Prog. Ser.* 163:133–43.

Katz, L. A., G. B. McManus, O.L.O. Snoeyenbos-West, et al. 2005. reframing the "Everything is everywhere" debate: Evidence for high gene flow and diversity in ciliate morphospecies. *Aquat. Microb. Ecol.* 41:55–65.

Kesseler, H. 1966. Beitrag zur Kenntnis der chemischen und physikalischen Eigenscahfetn del Zellsaftes von *Noctiluca miliaris*. *Veroff. Inst. Meeresforsch. Bremerh. Sonderband* 2:357–68.

Kiørboe, T. 1989. Growth in fish larvae: are they particularly efficient? *Rapp. P.-v. Réun. Cons. Int. Explor. Mer.* 191:383–89.

———. Turbulence, phytoplankton cell size and the structure of pelagic food webs. *Adv. Mar. Biol.* 29:1–72.

———. 1996. Material flux in the water column. In *Eutrophication in coastal marine ecosystems*, ed. B. B. Jørgensen and K. Richardson, 67–94. Coastal and Estuarine Studies, 52. Washington, DC: American Geophysical Union.

———. 1997. Small-scale turbulence, marine snow, and planktivorous feeding. *Sci. Mar.* 61(Suppl. 1):141–58.

———. 1998. Population regulation and role of mesozooplankton in shaping marine pelagic food webs. *Hydrobiologia* 363:13–27.

———. 2001. Formation and fate of marine snow: small-scale processes with large-scale implications. *Sci. Mar.* 65(Suppl. 2):57–71.

———. 2003. Marine snow microbial communities: scaling of abundances with aggregate size. *Aquat. Microb. Ecol.* 33:67–75.

———. 2006. Sex, sex ratios, and the dynamics of pelagic copepod populations. *Oecologia* 148:40–50.

———. 2007. Mate finding, mating, and population dynamics in a planktonic copepod *Oithona davisae*: There are too few males. *Limnol. Oceanogr.* 52:1511–22.

Kiørboe, T., K. P. Andersen, and H. Dam. 1990. Coagulation efficiency and aggregate formation in marine phytoplankton. *Mar. Biol.* 107:235–45.

Kiørboe, T., and E. Bagøien. 2005. Motility patterns and mate encounter rates in planktonic copepods. *Limnol. Oceanogr.* 50:1999–2007.

Kiørboe, T., H. P. Grossart, H. Ploug, et al. 2002. Bacterial colonization of sinking aggregates: mechanisms and rates. *Appl. Environ. Microbiol.* 68:3996–4006.

Kiørboe, T., and G. A. Jackson. 2001. Marine snow, organic solute plumes, and optimal chemosensory behaviour of bacteria. *Limnol. Oceanogr.* 46:1309–18.

Kiørboe, T., C. Lundsgaard, M. Olesen, et al. 1994. Aggregation and sedimentation processes during a spring phytoplankton bloom: a field experiment to test coagulation theory. *J. Mar. Res.* 52:297–323.

Kiørboe, T., F. Møhlenberg, and K. Hamburger. 1985. Bioenergetics of the planktonic copepod *Acartia tonsa*: relation between feeding, egg production and respiration, and the composition of specific dynamic action. *Mar. Ecol. Prog. Ser.* 26:85–95.

Kiørboe, T., and P. Munk. 1986. Feeding and growth of larval herring, *Clupea harengus*, in relation to density of copepod nauplii. *Env. Biol. Fishes* 17:133–39.

Kiørboe, T., P. Munk, and K. Richardson. 1987. Respiration and growth of larval herring *Clupea harengus*: Relation between specific dynamic action and growth efficiency. *Mar. Ecol. Prog. Ser.* 40:1–10.

Kiørboe, T., P. Munk, K. Richardson, V. Christensen, and H. Paulsen. 1988. Plankton dynamics and herring larval growth, drift and survival in a frontal area. *Mar. Ecol. Prog. Ser.* 44:205–19.

Kiørboe, T., and T. G. Nielsen. 1990. Effects of wind stress on vertical water column structure, phytoplankton growth, and fecundity of planktonic copepods. In *Trophic relationships in the marine environment,* ed. M. Barnes and R. N. Gibson. 28–40. Aberdeen: Aberdeen University Press.

Kiørboe, T., and T. G. Nielsen. 1994. Regulation of zooplankton biomass and production in a temperate, coastal ecosystem. I. Copepods. *Limnol. Oceanogr.* 39:493–507.

Kiørboe, T., H. Ploug, and U. H. Thygesen. 2001. Fluid motion and solute distribution around sinking aggregates. I. Small-scale fluxes and heterogeneity of nutrients in the pelagic environment. *Mar. Ecol. Prog. Ser.* 211:1–13.

Kiørboe, T., and M. Sabatini. 1994. Reproductive and life cycle strategies in egg-carrying cyclopoid and free-spawning calanoid copepods. *J. Plankton Res.* 16:1353–66.

———. 1995. The scaling of fecundity, growth and development in planktonic copepods. *Mar. Ecol. Prog. Ser.* 120:285–98.

Kiørboe, T., and E. Saiz. 1995. Planktivorous feeding in calm and turbulent environments, with emphasis on copepods. *Mar. Ecol. Prog. Ser.* 122:135–45.

Kiørboe, T., E. Saiz, and M. Viitasalo. 1996. Prey switching in the planktonic copepod *Acartia tonsa. Mar. Ecol. Prog. Ser.* 143:65–75.

Kiørboe, T., E. Saiz, and A. Visser. 1999. Hydrodynamic signal perception in the copepod *Acartia tonsa. Mar. Ecol. Prog. Ser.* 179:97–111.

Kiørboe, T., K. Tang, H. P. Grossart, et al. 2003. Microbial community dynamics on marine snow aggregates: colonization, growth, detachment and grazing mortality of attached bacteria. *Appl. Environ. Microbiol.* 69:3036–47.

Kiørboe, T., and U. H. Thygesen. 2001. Fluid motion and solute distribution around sinking aggregates. II. Implications for remote detection by colonizing zooplankters. *Mar. Ecol. Prog. Ser.* 211:15–25.

Kiørboe, T., and J. Titelman. 1998. Feeding, prey selection and prey encounter mechanisms in the heterotrophic dinoflagellate *Noctiluca scintillans. J. Plankton Res.* 20:1615–36.

Kiørboe, T., and A. W. Visser. 1999. Predator and prey perception in copepods due to hydromechanical signals. *Mar. Ecol. Prog. Ser.* 179:81–95.

Knauer, G. A., J. H. Martin, and K. W. Bruland. 1979. Fluxes of particulate carbon, nitrogen, and phosphorus in the upper water column of the northeast Pacific. *Deep-Sea Res.* 26A:97–108.

Koch, A. L. 1971. The adaptive responses of *Eschericia coli* to a feast and famine existence. *Adv. Micr. Physiol.* 6:147–217.

Koch, A. L. 1997. Microbial physiology and ecology of slow growth. *Molec. Biol. Rev.* 671:305–18.

Koehl, M.A.R., and J. R. Strickler. 1981. Copepod feeding currents: Food capture at low Reynolds number. *Limnol. Oceanogr.* 26:1062–73.

Koike, I., S. Hara, T. Terauchi, and K. Kogure. 1990. Role of sub-micrometer particles in the ocean. *Nature* 345:242–44.

Kokko, H., and B.B.M. Wong. 2007. What determines sex roles in mate searching? *Evolution* 61:1162–75.

Köster, F. W., and C. Möllmann. 2000. Egg cannibalism in Baltic sprat *Sprattus sprattus. Mar. Ecol. Prog. Ser.* 196:269–77.

Kozlowski, J. 1992. Optimal allocation of resources to growth and reproduction: Implications for age and size at maturity. *Trends in Ecol. and Evol.* 6:15-9.

Kunze, E., J. F. Dower, I. Beveridge, R. Dewey, and K. Bartlett. 2006. Observations of biologically generated turbulence in a coastal inlet. *Science* 313:1768-70.

Landry, M. R. 1981. Switching between herbivory and carnivory by the planktonic marine copepod *Calanus pacificus*. *Mar. Biol.* 65:77-82.

———. 1983. The development of marine calanoid copepods with comment on the isochronal rule. *Limnol. Oceanogr.* 28:614-24.

Landry, M. R., J. M. Lehner-Fournier, J. A. Sundstrom, V. L. Fagerness, and K. E. Selph. 1991. Discrimination between living and heat-killed prey by a marine zooflagellate, *Paraphysomonas vestita* (Stokes). *J. Exp. Mar. Biol. Ecol.* 146:139-51.

Lauria, M. L., D. A. Purdie, and J. Sharples. 1999. Contrasting phytoplankton distributions controlled by tidal turbulence in an estuary. *J. Mar. Systems* 21:189-97.

Lawrence, J. R., D. R. Korber, G. M. Wolfaardt, and D. E. Caldwell. 1995. Behavioral strategies of surface-colonizing bacteria. *Adv. Microb. Ecol.* 14:1-75.

Lawton, J. H., J. Beddington, and R. Bonser. 1974. Switching in invertebrate predators. In: *Ecological stability*, ed. M. B. Usher and M. H. Williamson. 141-58. London: Chapman and Hall.

Le Fèvre, L. 1986. Aspects of the biology of frontal systems. *Adv. Mar. Biol.* 23:163-299.

Legendre, L. 1981. Hydrodynamic control of marine phytoplankton production: The paradox of stability. In *Ecohydrodynamics*, ed. J.C.J. Nihoul, 197-207. Amsterdam: Elsevier.

———. 1990. The significance of microalgal blooms for fisheries and for the export of particulate organic carbon in the oceans. *J. Plankton Res.* 12:681-99.

Legendre, L., S. Demers, and D. LeFaivre. 1986. Biological production at marine ergoclines. In *Marine interfaces ecohydrodynamics*, ed. J.C.J. Nihoul, 1-29. Amsterdam: Elsevier.

Legendre, L., and J. Le Fèvre. 1989. Hydrodynamical singularities as controls of export production in the oceans. In *Productivity of the ocean: Present and past*, ed. W. H. Berger, V. S. Smetacek, and G. Wefer. 44-63. Chichester, UK: John Wiley & Sons.

Levandowsky, M., B. S. White, and F. L. Schuster. 1997. Random movements of soil amebas. *Acta Protozoologica* 36:237-48.

Lewis, D. M., and T. J. Pedley. 2001. The influence of turbulence on plankton predation strategies. *J. Theor. Biol.* 210:347-65.

López-Garcia, P., F. Rodriguez-Valera, C. Pedrós-Alió, and D. Moreira. 2001. Unexpected diversity of small eukaryotes in deep-sea Antarctic plankton. *Nature* 409:603-7.

Mackas, D. L., H. Sefton, C. B. Miller, et al. 1993. Vertical habitat partitioning by large calanoid copepods in the oceanic sub-arctic pacific during spring. *Prog. Oceanogr.* 32:259-94.

MacKenzie, B. R., and T. Kiørboe. 1995. Encounter and swimming behaviour of pause-travel and cruise larval fish predators in calm and turbulent laboratory environments. *Limnol. Oceanogr.* 40:1278-89.

———. 2000. Larval fish feeding in turbulence: A case for the downside *Limnol. Oceanogr.* 405:1–10.

MacKenzie, B. R., T. J. Miller, S. Cyr, and W. C. Leggett. 1994. Evidence for a dome-shaped relationship between turbulence and larval fish contact rates. *Limnol. Oceanogr.* 39:1790–99.

Magar, V., T. Goto, and T. J. Pedley. 2003. Nutrient uptake by a self-propelled steady squirmer. *Q. J. Mech. Appl. Math.* 56:65–91.

Magar, V., and T. J. Pedley. 2005. Average nutrient uptake by a self-propelled unsteady squirmer. *J. Fluid Mech.* 539:93–112.

Malone, T. C. 1980. Algal size. In *The physiological ecology of phytoplankton*, ed. I. Morrison. 433–63. Oxford: Blackwell Scientific Publications.

Malone, T. C., E. P. Sharon, and D. J. Conley. 1993. Transient variations in phytoplankton productivity at the JGOFS Bermuda time series station. *Deep-Sea Res.* 40:903–24.

Mann, K. H., and J.R.N. Lazier. 1991. *Dynamics of marine ecosystems: Biological-physical interactions in the oceans.* Cambridge: Blackwell Scientific Publications.

Margalef, R. 1978. Life-forms of phytoplankton as survival alternatives in an unstable environment. *Oceanologica Acta* 1:493–509.

Marra, J., R. R. Bidigare, and T. D. Dickey. 1990. Nutrients and mixing, chlorophyll and phytoplankton growth. *Deep-Sea Res.* 37:127–43.

Marrase, C., J. H. Costello, T. Granata, and J. R. Strickler. 1990. Grazing in a turbulent environment—energy dissipation, encounter rates, and efficacy of feeding currents in *Centropages hamatus*. *Proc. Natl. Acad. Sci. USA.* 87:1653–7.

Martin, J. H. 1990. Glacial-interglacial CO_2 changes: The iron hypothesis. *Palaeoceanography* 5:1–13.

Martin, J. H., K. H. Coale, K. S. Johnson, et al. 1994. Testing the iron hypothesis in ecosystems of the equatorial Pacific Ocean. *Nature* 371:123–9.

Mauchline, J. 1998. The biology of calanoid copepods. *Adv. Mar. Biol.* 33:1–710.

McGurk, M. D. 1986. Natural mortality of marine pelagic fish eggs and larvae—role of spatial pachiness. *Mar. Ecol. Prog. Ser.* 34:227–42.

Menden-Deuer, S., and D. Grünbaum. 2006. Individual foraging behaviors and population distributions of a planktonic predator aggregating to phytoplankton thin layers. *Limnol. Oceanogr.* 51:109–16.

Menden-Deuer, S., and E. J. Lessard. 2000. Carbon to volume relationships for dinoflagellates, diatoms, and other protist plankton. *Limnol. Oceanogr.* 45:569–79.

Menden-Deuer, S., E. J. Lessard, J. Satterberg, et al. 2005. Growth rates and starvation survival of three species of the pallium-feeding, thecate dinoflagellate genus *Protoperidinium*. *Aquat. Microb. Ecol.* 41:145–52.

Milne, A. A. 1926. *Winnie the Pooh.* London: Methuen.

Miquel, J. C., S. W. Fowler, J. La Rosa, and P. Buat-Menard. 1994. Dynamics of the downward flux of particles and carbon in the open northwestern Mediterranean Sea. *Deep-Sea Res. I.* 41:243–61.

Mitchell, J. G., L. Pearson, and S. Dillon. 1996. Clustering of marine bacteria in seawater enrichments. *Appl. Env. Microbiol.* 62:3716–21.

Moore, J. K., and T. A. Villareal. 1996. Size–ascent rate relationship in positively buoyant marine diatoms. *Limnol. Oceanogr.* 41:1514–20.

Moore, P. A., D. M. Fields, and J. Yen. 1999. Physical constraints of chemoreception in foraging copepods. *Limnol. Oceanogr.* 44:166–77.

Mopper, K., and P. Lindroth. 1982. Diel and depth variation in dissolved free amino acids and ammonium in the Baltic Sea determined by shipboard HPLC analysis. *Limnol. Oceanogr.* 27:336–47.

Mouritsen L. T., and K. Richardson. 2003. Vertical microscale patchiness in nano- and microplankton distributions in a stratified estuary. *J. Plankton Res.* 25:783–97.

Mullin, M. M., E. F. Stewart, and F. J. Fuglister. 1975. Ingestion by planktonic grazers as a function of concentration of food. *Limnol. Oceanogr.* 20:259–62.

Munk, P. 1992. Foraging behaviour and prey size spectra of larval herring *Clupea harengus. Mar. Ecol. Prog. Ser.* 80:149–58.

———. 1997. Prey size spectra and prey availability of larval and small juvenile cod. *J. Fish Biol.* 51 (Suppl. A):340–51.

———. 2007. Cross-frontal variation in growth rate and prey availability of larval North Sea cod *Gadus morhua. Mar. Ecol. Prog. Ser.* 334:225–35.

Munk, P., and T. Kiørboe. 1985. Feeding behaviour and swimming activity of larval herring (*Clupea harengus* L.) in relation to density of copepod nauplii. *Mar. Ecol. Prog. Ser.* 24:15–21.

Munk, P., P. O. Larsson, D. S. Danielssen, and E. Moksness. 1999. Variability in frontal zone formation and distribution of gadoid fish larvae at the shelf break in the northeastern North Sea. *Mar. Ecol. Prog. Ser.* 177:221–33.

Munk, W. H., and G. A. Riley. 1952. Absorption of nutrients by aquatic plants. *J. Mar. Res.* 11:215–40.

Munk, W. H., and G. A. Riley. 1953. Absorption of nutrients by aquatic plants. *J. Mar. Res.* 11:215–40.

Murdoch, W. W., and A. Oaten. 1975. Predation and population stability. *Adv. Ecol. Res.* 9:2–131.

Nair, R. R., V. Ittekkot, S. J. Manganini, V. Ramaswamy, B. Haake, E. T. Degens, B. N. Desai, and S. Honjo. 1989. Increased particle-flux to the deep ocean related to monsoons. *Nature* 338:749–51.

Nakamura, Y. 1998. Growth and grazing of the large heterotrophic dinoflagellates, *Noctiluca scintillans*, in laboratory cultures. *J. Plankton Res.* 20:1711–20.

Naustvoll, L. J. 2000. Prey size spectra in naked heterotrophic dinoflagellates. *Phycologia* 39:448–55.

Nielsen, E., and K. Richardson. 1996. Can changes in the fisheries yield in the Kattegat (1950–1992) be linked to changes in primary production? *ICES J. Mar. Sci.* 53:988–94.

Nielsen, S. L., S. Enriquez, C. M. Duarte, and K. Sand-Jensen. 1996. Scaling maximum growth rates across photosynthetic organisms. *Funct. Ecol.* 10:167–75.

Nielsen, T. G. 2005. *Struktur og funktion af fødenettet i havets frie vandmasser.* Roskilde, Denmark: Danmarks Miljøundersøgelser.

Nielsen, T. G., B. Løkkegaard, K. Richardson, F. B. Pedersen, and L. Hansen. 1993. Structure of plankton communities in the Dogger Bank area (North Sea) during a stratified situation. *Mar. Ecol. Prog. Ser.* 95:115-31.

Nishii, S. 1998. Hydromechanical perception in chaetognaths (in Japanese—read to me by Hiroaki Saito). MSc thesis, Mie University.

Nixon, S. W. 1992. Quantifying the relationship between nitrogen input and the productivity of marine ecosystems. *Proc. Adv. Mar. Tech. Conf.* 5:57-83.

Ohman, M. D., B. W. Frost, and E. B. Cohen. 1983. Reverse diel vertical migration—an escape from invertebrate predators. *Science* 220:1404-7.

Okubo, A. 1980. *Diffusion and ecological problems: Mathematical models.* Berlin: Springer.

Olsen, E. M., M. Heino, G. R. Lilly, M. J. Morgan, J. Brattey, B. Ernando, and U. Dieckmann. 2004. Maturation trends indicative of rapid evolution precedes the collapse of northern cod. *Nature* 428:932-35.

Olsen, Y. 2002. MARICULT Research Programme: Background, status and main conclusions. *Hydrobiologia* 484:1-10.

Osborn, T. 1996. The role of turbulent diffusion for copepods with feeding currents. *J. Plankton Res.* 18:185-95.

Paffenhöfer, G. A. 1993. On the ecology of marine cyclopoid copepods (Crustacea: Copepoda, Cyclopoida). *J. Plankton Res.* 15:37-55.

Parrish, K. K., and D. F. Wilson. 1978. Fecundity studies on *Acartia tonsa* (Copepoda, Calanoida) in standardized culture. *Mar. Biol.* 46:65-81.

Pauly, D., and V. Christensen. 1995. Primary production required to sustain global fisheries. *Nature* 374:255-57.

Pauly, D., V. Christensen, J. Dalsgaard, R. Froese, and F. Torres, Jr. 1998. Fishing down the marine food webs. *Science* 279:860-3.

Penry, D. L., and P. A. Jumars. 1986. Chemical reactor analysis and optimal digestion. *Bioscience* 36:310-13.

Peters, R. H., E. Demers, M. Koelle, and B. R. MacKenzie, 1994. The allometry of swimming speed and predation. *Verh. Int. Verein. Limnol.* 25:2316-23.

Peterson, W. T., D. F. Arcos, G. B. McManus, H. Dam, D. Bellantoni, T. Johnson, and P. Tiselius. 1988. The nearshore zone during coastal upwelling—daily variability and coupling between primary and secondary production off central Chile. *Prog. Oceanogr.* 20:1-40.

Peterson, W. T., and W. J. Kimmerer. 1994. Processes controlling recruitment of the marine calanoid copepod *Temora longicornis* in Long Island Sound: Egg production, egg mortality, and cohort survival rates. *Limnol. Oceanogr.* 39:1594-605.

Pielou, E. C. 1969. *An Introduction to Mathematical Ecology.* New York: Wiley-Interscience.

Platt, T. 1985. Structure of marine ecosystems: its allometric basis. *Can. Bull. Fish. Aquat. Sci.* 213:55-64.

Ploug, H., and H.-P. Grossart. 1999. Bacterial production and respiration in suspended aggregates—a matter of the incubation method. *Aquat. Microb. Ecol.* 20:21-29.

Ploug, H., and U. Passow. 2007. Direct measurements of diffusion within diatom aggregates containing transparent exopolymeric particles. *Limnol. Oceanogr.* 52:1-6.

Ploug, H., W. Stolte, and B. B. Jørgensen. 1999. Diffusive boundary layers of the colony-forming plankton alga *Phaeocystis* sp.—implications for nutrient uptake and cellular growth. *Limnol. Oceanogr.* 44:1959–67.

Pomeroy, L. R. 1974. Oceans food web, a changing paradigm. *Bioscience* 9:499–504.

Pommier, T., J. Pinhassi, and Å. Hagström. 2005. Biogeographic analysis of ribosomal RNA clusters from marine bacterioplankton. *Aquat. Microb. Ecol.* 41:79–89.

Poulet, S. A., R. Williams, D.V.P. Conway, and C. Videau. 1991. Co-occurrence of copepods and dissolved free amino acids in shelf waters. *Mar. Biol.* 108:373–85.

Price, H. J., and G.-A. Paffenhöfer. 1986. Effect of concentration of the feeding of a marine copepod in algal monocultures and mixtures. *J. Plankton Res.* 8:119–28.

Pruppacher, H. R., and J. D. Klett. 1978. *Microphysics of clouds and precipitation.* Norwell, MA: D. Reidel.

Purcell, E. M. 1977. Life at low Reynolds number. *Am. J. Phys.* 45:3–11.

Raven, J. A. 1997. The vacuole: A cost-benefit analysis. *Adv. Bot. Res.* 25:59–86.

Raven, J. A., and K. Richardson. 1984. Dinophyte flagella—a cost benefit analysis. *New Phytol.* 98:259–76.

Richardson, A. J., H. M. Verheye, B. A. Mitchell-Innes, J. L. Fowler, and J. G. Field. 2003. Seasonal and event-scale variation in growth of *Calanus agulhensis* (Copepoda) in the Benguela upwelling system and implications for spawning of sardine *Sardinops sagax*. *Mar. Ecol. Prog. Ser.* 254:239–51.

Richardson, K., M. F. Lavín-Peregrina, E. G. Mitchelson, and J. H. Simpson. 1985. Seasonal distribution of chlorophyll a in relation to physical structure in the Western Irish Sea. *Oceanologica Acta* 8:77–86.

Richardson, K., T. G. Nielsen, F. B. Pedersen, et al. 1998. Spatial heterogeneity in the structure of the planktonic food web in the North Sea. *Mar. Ecol. Prog. Ser.* 168:197–211.

Richerson, P., R. Armstrong, and C. R. Goldman. 1970. Contemporaneous disequilibrium, a new hypothesis to explain the "Paradox of the Plankton." *Proc. Natl. Acad. Sci. USA* 67:1710–4.

Riebesell, U., M. Reigstad, P. Wassmann, T. Noji, and U. Passow. 1995. On the trophic fate of *Phaeocystis pouchetii* (Hariot): VI. Significance of *Phaeocystis*-derived mucus for vertical flux. *Neth. J. Sea Res.* 33:193–203.

Riemann, L., G. F. Steward, and F. Azam. 2000. Dynamics of bacterial community composition and activity during a mesocosm diatom bloom. *Appl. Environ. Microbiol.* 66:578–87.

Riemann L., and A. Winding. 2001. Community dynamics of free-living and particle-associated bacterial assemblages during a freshwater phytoplankton bloom. *Microb. Ecol.* 42:274–85.

Riley, G. A., H. Stommel, and D. F. Bumpus. 1949. Quantitative ecology of the plankton of the western North Atlantic. *Bull. Bingham Oceananographic Collection Yale University* 12:1–169.

Rodriguez, J., J. Tintoré, J. T. Allen, J. M. Blanco, D. Gomis, A. Reul, J. Ruiz, V. Rodriguez, F. Echevarria, and F. Jiménez-Gomez. 2001. Mesoscale vertical motion and the size structure of phytoplankton in the Ocean. *Nature* 410:360–63.

Rothschild, B. J., and T. R. Osborn. 1988. Small-scale turbulence and plankton contact rates. *J. Plankton Res.* 10:465–74.

Roy, S., N. Silverberg, N. Romero, D. Deibel, B. Klein, C. Savenkoff, A. F. Vézina, J.-É. Tremblay, L. Legendre, and R. B. Rivkin. 2000. Importance of mesozooplankton feeding for the downward flux of biogenic carbon in the gulf of St. Lawrence (Canada). *Deep-Sea Res II.* 47:519–44.

Runge, J. A., and S. Plourde. 1996. Fecundity characteristics of *Calanus finmarchicus* in coastal waters of eastern Canada. *Ophelia* 44:171–87.

Ryther, J. H. 1969. Photosynthesis and fish production in the sea. *Science* 166:72–76.

Saiz, E., A. Calbet, and E. Broglio. 2003. Effects of small-scale turbulence on copepods: The case of *Oithona davisae. Limnol. Oceanogr.* 48:1304–11.

Saiz, E., and T. Kiørboe. 1995. Predatory and suspension-feeding of the copepod *Acartia tonsa* in turbulent environments. *Mar. Ecol. Prog. Ser.* 122:147–58.

Sampei, M., H. Sasaki, H. Hattori, S. Kudoh, Y. Kashino, and M. Fukuchi. 2002. Seasonal and spatial variability in the flux of biogenic particles in the North Water, 1997–1998. *Deep-Sea Res. II.* 49:5245–57.

Sandstrom, O. 1980. Selective feeding by Baltic herring. *Hydrobiologia* 69:199–207.

Sarmiento, J. L., and J. C. Orr. 1991. Three-dimensional simulations of the impact of Southern Ocean nutrient depletion on atmospheric CO_2 and ocean chemistry. *Limnol. Oceanogr.* 36:1928–50.

Selander, E., P. Thor, G. Toth, and H. Pavia. 2006. Copepods induce paralytic shellfish toxin production in marine dinoflagellates. *Proc. R. Soc. Biol. Sci.* 273:1673–80.

Seymour, J. R., J. G. Mitchell, L. Pearson, and R. L. Waters. 2000. Heterogeneity in bacterioplankton abundance from 4.5 millimetre resolution sampling. *Aquat. Microb. Ecol.* 22:142–53.

Sharples, J., J. F. Tweddle, J. A. Green, et al. 2007. Spring-neap modulation of internal tide mixing and vertical nitrate fluxes at a shelf edge in summers. *Limnol. Oceanogr.* 52:1735–47.

Sheldon, R. W., A. Prakash, and W. H. Sutcliffe, Jr. 1972. The size distribution of particles in the ocean. *Limnol. Oceanogr.* 17:327–40.

Shimeta, J. 1993. Diffusional encounter of submicron particles and small cells by suspension feeders *Limnol. Oceanogr.* 38:456–65.

Shimeta, J., P. A. Jumars, and E. J. Lessard. 1995. Influence of turbulence on suspension feeding by planktonic protozoa: experiments in laminar shear fields. *Limnol. Oceanogr.* 40:845–59.

Silver, M. W., and M. M. Gowing. 1991. The "particle" flux: Origins and biological components. *Prog. Oceanogr.* 26:75–113.

Simon, M., H.-P. Grossart, B. Schweitzer, and H. Ploug. 2002. Microbial ecology of organic aggregates in aquatic systems. *Aquat. Microb. Ecol.* 28:175–211.

Simpson, J. H., and J. R. Hunter. 1974. Fronts in Irish Sea. *Nature* 250:404–6.

Singarajah, K. V. 1969. Escape reactions of zooplankton: the avoidance of a pursuing siphon tube. *J. Exp. Mar. Biol. Ecol.* 3:171–78.

Smayda, T. J. 1970. The suspension and sinking of phytoplankton in the sea. *Oceanogr. Mar. Biol. Ann. Rev.* 8:353–414.

Smetacek, V. 2001. A watery arms race. *Nature* 411:745.

Smetacek, V., P. Assmy, and J. Henjes. 2004. The role of grazing in structuring Southern Ocean pelagic ecosystems and biogeochemical cycles. *Antarctic Science* 16:541–58.

Smetacek, V., K. V. Brockel, B. Zeitzschel, and W. Zenk. 1978. Sedimentation of particulate matter during a phytoplankton spring bloom in relation to hydrographical regime. *Mar. Biol.* 47:211–26.

Smetacek, V., and F. Pollehne. 1986. Nutrient cycling in pelagic systems—a reappraisal of the conceptual-framework. *Ophelia* 26:401–28.

Smith, D. C., M. Simon, A. L. Alldredge, and F. Azam. 1992. Intensive hydrolytic activity on marine aggregates and implications for rapid particle dissolution. *Nature* 359:139–41.

Sommer, U. 1983. Nutrient composition between phytoplankton species in multispecies chemostat experiments. *Arch. Hydrobiol.* 96:399–416.

———. 1984. The paradox of the plankton—fluctuations of phosphorus availability maintain diversity of phytoplankton in flow-through cultures. *Limnol. Oceanogr.* 29:633–36.

———. 1988. Some size relationships in phytoflagellate motility. *Hydrobiologia* 161:125–31.

Sournia, A., J.-L. Birrien, J.-L. Douvillé, B. Klein, and M. Viollier. 1987. A daily study of the diatom spring bloom at Roscoff (France) in 1985. I. The spring bloom within the annual cycle. *Est. Coast. Shelf Sci.* 25:355–67.

Spehr, M., G. Gisselmann, A. Poplawski, J. A. Riffell, C. H. Wetzel, R. H. Zimmer, and H. Hatt. 2003. Identification of a testicular odorant receptor mediating human sperm chemotaxis, *Science* 299:2054–58.

Stearns, S. C. 1992. *The evolution of life histories.* Oxford: Oxford University Press.

Steele, J. H. 1974. *The structure of marine ecosystems.* Cambridge, MA: Harvard University Press.

———. 1977. Structure of plankton communities. *Phil. Trans. R. Soc. Lond. B* 280:485–534.

———. 1978. Some comments on plankton patches. In *Spatial patterns in plankton communities,* ed. J. H. Steele. New York: Plenum Press.

Steinberger, R. E., A. R. Alle, H. G. Hansma, and P. A. Holden. 2002. Elongation correlates with nutrient deprivation in *Pseudomonas aeruginosa*-unsaturated biofilm. *Microb. Ecol.* 43:416–23.

Stephens, P. A., W. J. Sutherland, and R. P. Freckleton. 1999. What is the Allee effect? *Oikos* 87:185–90.

Stoecker, D. 1998. Conceptual models of mixotrophy in planktonic protists and some ecological and evolutionary implications. *Eur. J. Protist.* 34:281–90.

Strickler, J. R., and A. K. Bal. 1973. Setae of the first antennae of the copepod *Cyclops scutifer* (Sars): Their structure and importance. *Proc. Natl. Acad. Sci. USA.* 70:2656–59.

Strom, S. L., and E. D. Buskey. 1993. Feeding, growth, and behaviour of the thecate heterotrophic dinoflagellate *Oblea rotunda. Limnol. Oceanogr.* 38:965–77.

Suttle, C. A., A. M. Chan, and J. A. Fuhrman. 1991. Dissolved free amino acids in the Sargasso Sea: Uptake and respiration rates, turnover times, and concentrations. *Mar. Ecol. Prog. Ser.* 70:189–99.

Svensen, C., and T. Kiørboe. 2000. Remote prey detection in *Oithona similis*: Hydromechanical versus chemical cues. *J. Plankton Res.* 22:1155–66.

Sverdrup, H. U. 1953. On conditions for the vernal blooming of phytoplankton. *J. Cons. Int. Expl. Mer.* 18:287–95.

Taylor, G. T. 1989. Variability in the vertical flux of microorganisms and biogenic material in the epipelagic zone of a North Pacific central gyre station. *Deep-Sea Res.* 36:1287–308.

Teegarden, G. J. 1999. Copepod grazing selection and particle discrimination on the basis of PSP toxin content. *Mar. Ecol. Prog. Ser.* 181:163–76.

Thingstad, T. F. 1998. A theoretical approach to structuring mechanisms in the pelagic food web. *Hydrobiologia* 363:59–72.

———. 2000. Elements of a theory for the mechanisms controlling abundance, diversity, and biogeochemical role of lytic bacterial viruses in aquatic systems. *Limnol. Oceanogr.* 45:1320–28.

Thingstad, T. F., H. Havskum, K. Garde, and B. Riemann. 1996. On the strategy of "eating your competitor": A mathematical analysis of algal mixotrophy. *Ecology* 77:2108–18.

Thingstad, T. F., L. Øvreås, J. K. Egge, T. Løvdal, and M. Heldal. 2005. Use of non-limiting substrates to increase size; a generic strategy to simultaneously optimize uptake and minimize predation in pelagic osmotrophs. *Ecol. Letters* 8:675–82.

Thomsen, H. A., G. Hansen, J. Larsen, Ø. Moestrup, and N. Vørs. 1992. Fytoplankton og heterotroft nanoplankton. In *Planktondynamik og stofomsœtning i Kattegat*, ed. T. Fenchel. 31–59. Copenhagen: Miljøstyrelsen, Danish Ministry of Environment.

Thor, P. 2000. Relationship between specific dynamic action and protein deposition in calanoid copepods. *J. Exp. Mar. Biol. Ecol.* 245:171–82.

Tiselius, P. 1992. Behavior of *Acartia tonsa* in patchy food environments. *Limnol. Oceanogr.* 37:1640–51.

———. 1998. Short term feeding responses to starvation in three species of small calanoid copepods. *Mar. Ecol. Prog. Ser.* 168:119–26.

Tiselius, P., and P. Jonsson. 1990. Foraging behaviour of six calanoid copepods: Observations and hydrodynamic analysis. *Mar. Ecol. Prog. Ser.* 66:23–33.

Titelman, J. 2001. Swimming and escape behaviour of copepod nauplii: Implications for predator-prey interactions among copepods. *Mar. Ecol. Prog. Ser.* 213:203–13.

Titelman, J., and O. Fiksen. 2004. Ontogenetic vertical distribution patterns in small copepods: field observations and model predictions. *Mar. Ecol. Prog. Ser.* 284:49–63.

Titelman, J., and T. Kiørboe. 2003. Motility of copepod nauplii and implications for food encounter. *Mar. Ecol. Prog. Ser.* 247:123–35.

Tranvik, C. J., E. B. Sherr, and B. F. Sherr. 1993. Uptake and utilization of "colloidal DOM" by heterotrophic flagellates in seawater. *Mar. Ecol. Prog. Ser.* 92:301–9.

Tsuda, A., and C. B. Miller. 1998. Mate-finding behaviour in *Calanus marshallae* Frost. *Phil. Trans. R. Soc. B* 353:713–20.

Uchima, M., and R. Hirano. 1986. Food of *Oithona davisae* (Copepoda: Cyclopoida) and the effect of food concentration at first feeding on the larval growth. *Bull. Plankton Soc. Jpn.* 33:21–28.

Urban-Rich, J., E. Nordby, I. J. Andreassen, and P. Wassmann. 1999. Contribution by mesozooplankton fecal pellets to the carbon flux on Nordvestbanken, north Norwegian shelf in 1994. *Sarsia* 84:253–64.

Urrère, M. A., and G. A. Knauer. 1981. Zooplankton fecal pellet fluxes and vertical transport of particulate organic material in the pelagic environment. *J. Plankton Res.* 3:369–87.

van Duren, L. A., E. J. Stamhuis, and J. J. Videler. 1998. Reading the copepod personal ads: increasing encounter probability with hydromechanical signals. *Phil. Trans. R. Soc. B* 353:691–700.

Venter, J. C., K. Remington, J. F. Heidelberg, et al. 2004. Environmental genome shotgun sequencing of the Sargasso Sea. *Science* 304:66–74.

Verity, P. G. 1991. Measurement and simulation of prey uptake by marine planktonic ciliates fed plastidic and aplastidic nanoplankton. *Limnol. Oceanogr.* 36:729–50.

Verity, P. G., and V. Smetacek. 1996. Organism life cycles, predation, and the structure of marine pelagic ecosystems. *Mar. Ecol. Prog. Ser.* 130:277–93.

Viitasalo, M., T. Kiørboe, J. Flinkman, L. W. Pedersen, and A. W. Visser. 1998. Predation vulnerability of planktonic copepods: consequences of predator foraging strategies and prey sensory abilities. *Mar. Ecol. Prog. Ser.* 175:129–42.

Villareal, T. A. 1988. Positive buoyancy in the oceanic diatom *Rhizosolenia debyana* H. Peragallo. *Deep-Sea Res.* 35:1037–45.

Villareal, T. A., M. A. Altabet, and K. Culver-Rymsza. 1993. Nitrogen transport by vertically migrating diatom mats in the North Pacific. *Nature* 363:709–12.

Visser, A. W. 2001. Hydromechanical signals in the plankton. *Mar. Ecol. Prog. Ser.* 222:1–24.

———. 2007. Biomixing of the ocean. *Science* 316:838–39.

Visser, A. W., and G. A. Jackson. 2004. Characteristics of the chemical plume behind a sinking particle in a turbulent water column. *Mar. Ecol. Prog. Ser.* 283:55–71.

Visser, A. W., and T. Kiørboe. 2006. Plankton motility patterns and encounter rates. *Oecologia* 148:538–46.

Visser, A. W., and U. H. Thygesen. 2003. Random motility of plankton: Diffusive and aggregative contributions. *J. Plankton Res.* 25:1157–68.

Viswanathan, G. M., V. Afanasyev, S. V. Buldyrev, E. J. Murphy, P. A. Prince, and H. E. Stanley. 1996. Lévy flight search patterns of wandering albatrosses. *Nature* 381:413–15

Vogel, S. 1994. *Life in moving fluids. The physical biology of flow.* 2nd ed. Princeton: Princeton University Press.

Vuorinen, I. 1987. Vertical migration of *Eurytemora* (Crustacea, Copepoda): A compromise between the risks of predation and decreased fecundity. *J. Plankton Res.* 9:1037–46.

Waggett R., and J. H. Costello. 1999. Capture mechanisms used by the lobate ctenophore, *Mnemiopsis leidyi*, preying on the copepod *Acartia tonsa*. *J. Plankton Res.* 21:2037–52.

Walsby, A. E., P. K. Hayes, R. Boje, and L. J. Stal. 1997. The selective advantage of buoyancy provided by gas vesicles for planktonic cyanobacteria in the Baltic Sea. *New Phytologist* 136:407–17.

Ward, B. B. 2000. Nitrification and the marine nitrogen cycle. In *Microbial ecology of the oceans*, ed. D. L. Kirchman. 427–53. New York: Wiley-Liss.

Ward, G. E., C. J. Brokaw, D. L. Garbers, and V. D. Vacquier, 1985. Chemotaxis of *Arbacia punctulata* spermatozoa to resact, a peptide from the egg jelly layer. *J. Cell Biol.* 101:2324–29.

Ward, T. M., L. J. McLeay, W. F. Dimmlich, P. J. Rogers, S.A.M. McClatchie, R. Matthews, J. Kampf, and P. D. van Ruth. 2006. Pelagic ecology of a northern boundary current system: effects of upwelling on the production and distribution of sardine (*Sardinops sagax*), anchovy (*Engraulis australis*) and southern bluefin tuna (*Thunnus maccoyii*) in the Great Australian Bight. *Fish. Oceanogr.* 15:191–207.

Wassmann, P. 1990. Relationship between primary and export production in the boreal coastal zone of the North Atlantic. *Limnol. Oceanogr.* 35:464–71.

Waters, R. L., J. G. Mitchell, and J. Seymour. 2003. Geostatistical characterisation of centimetre-scale spatial structure of in vivo fluorescence. *Mar. Ecol. Prog. Ser.* 251:49–58.

Wells, M., and E. D. Goldberg. 1991. Occurrence of small colloids in sea water. *Nature* 353:342–44.

Wexels Riser, C. 1996. Eksport og retensjon av organisk materiale i en nordnorsk fjord (Balsfjorden, 1996): betydningen av mesozooplankton og krill. MSc thesis. Tromsø, Norway: University of Tromsø.

Wexels Riser, C., P. Wassmann, K. Olli, A. Pasternak, and E. Arashkevich. 2002. Seasonal variation in production, retention and export of zooplankton faecal pellets in the marginal ice zone and central Barents Sea. *J. Mar. Syst.* 38:175–88.

Yen, J., and D. M. Fields. 1992. Escape responses of *Acartia hudsonica* nauplii from the flow field of *Temora longicornis*. *Arch. Hydrobiol. Beih.* 36:123–34.

Yen, J., P. H. Lenz, D. V. Gassie, and D. K. Hartline. 1992. Mechanoreception in marine copepods: electrophysiological studies on the first antennae. *J. Plankton Res.* 14:459–512.

Yen, J., and J. R. Strickler. 1996. Advertisement and concealment in the plankton: what makes a copepod hydrodynamically conspicuous. *Invertebr. Biol.* 115:191–205.

Yen, J., M. J. Weissburg, and M. H. Doall. 1998. The fluid physics of signal perception by mate-tracking copepods. *Phil. Trans. R. Soc. B* 353:787–804.

Zeldis, J. R., J. Oldman, S. L. Ballara, and L. A. Richards. 2005. Physical fluxes, pelagic ecosystem structure, and larval fish survival in Hauraki Gulf, New Zealand. *Can. J. Fish. Aquat. Sci.* 62:593–610.

Zimmer-Faust, R. K., and M. Tamburi. 1994. Chemical identity and ecological implications of a waterborne settlement cue. *Limnol. Oceanogr.* 39.1075–87.

INDEX